WONDERMENT

SCIENCE IN STORIES

ROGER KENYON

for my brother, Richard

— survivor and smartest guy I know

It would be possible to describe everything scientifically, but it would make no sense; it would be without meaning, as if you described a Beethoven symphony as a variation of wave pressure.

ALBERT EINSTEIN

No, no! The adventures first, explanations take such a dreadful time.

LEWIS CARROLL

INTRODUCTION

The world works in wonderful ways. Appreciating them is what Wonderment is all about. But this isn't a textbook. There are no calculations or quizzes. No vocabulary lists to memorize.

There are stories: episodes of science in the everyday. In the kitchen, at a cabin, at a garage sale, in dirty laundry. Stories are familiar and approachable and make big ideas easier to remember.

Big ideas in science are patterns. For instance, objects are lazy by nature. Inertia. Things wear down unless kept up. Entropy. Wonderment explores these patterns in stories to focus on the big picture. To appreciate how life happens and, in doing so, see the familiar as if for the first time.

BIG IDEAS

Warm up to Winter: Heat and Energy

1. heat expands most matter
2. cold makes water expand
3. cold is the absence of heat
4. conduction can feel cool
5. heat flow can be slowed
6. liquids can reduce friction
7. insulation reduces heat flow
8. conduction is heat by touch
9. compression warms; expansion cools

Spring into Action: Matter and Motion

1. objects are lazy by nature
2. opposing contact creates friction
3. more time means less impact
4. solids have stability
5. faster fluid, lower pressure
6. smart bodies self-regulate
7. reaction is equal and opposite

8. soft bubbles and crunchy crystals
9. liquid surfaces have tension

Summer Sights and Sounds: Light and Sound

1. sound is vibration heard
2. speed affects sound perception
3. frequency varies with volume
4. frequency varies with motion
5. radio waves are vibrating electrons
6. colour varies with light frequency
7. colour reflected is colour detected
8. light and sound are forms of energy
9. light and sound can be absorbed or reflected

Eclectic Autumn: Electricity and Entropy

1. static electricity is built-up charge
2. like poles repel; opposites attract
3. magnetic attraction has practical uses
4. electricity flows through a conductor
5. energy changes, becoming less usable
6. an electric motor uses magnetism
7. things wear down unless kept up
8. resistance acts to oppose flow
9. measurement affects what is measured

Beyond the Year: Ideas on the Whole

1. nature is never ending
2. nature is ever emerging

CAST

Ajay and Jaya, neighbours

Alex (nephew of Ajay) and Lexa, spouses

Alan and Lana, children of Alex and Lexa

Clay and Lacy, friends of Alan

Mona (Clay's mom) and Noam (Lacy's dad), life partners

Karl and Lark, neighbours of Noam and Mona

PART 1: WARM UP TO WINTER

Heat and Energy

CHAPTER 1
HEAT EXPANDS MOST MATTER

Lightning Strikes, Wheel Washers, and Cup Cracks

Yard work can be sweaty work, even in autumn. Ajay takes a short break and sits under an apple tree. When warm, we relax. We stretch out and matter does too. Heat makes matter stretch out as its bits move apart.

Ajay sits in the shade, with an eye on clouds gathering over the horizon. Storm clouds could bring lightning. Rapid heat makes matter expand quickly, and a lightning stroke is both rapid and hot. Hotter than the surface of the sun. A bolt can blow your clothes off—or worse.

Lightning sends a shockwave of thunder as heat expands air around the bolt. Rapidly heated air spreads out as a shockwave that rumbles, echoing off the ground and off clouds.

Lightning striking a tree can travel through the sap, heating and expanding so fast it blows the bark off. People don't have sap, but we sweat. Doing yard work, Ajay has a layer of moisture over his skin. Lightning can heat and expand sweat like sap, turning it to steam so fast it blows off clothes.

LET'S leave Ajay under the tree and step inside. Loosening a lid is like a lightning strike, but a lot safer. To loosen a stubborn metal lid, run the jar under hot water. Heat in the water works like lightning, at least in concept. It expands metal in the lid, creating a gap between the jar and the lid stuck on by friction. Increasing the gap decreases friction, so the lid is easier to remove.

Pioneers attached wagon wheels this way, by heating a washer (a metal disc, like a coin). Unlike a coin, a washer has a hole in the centre. When heated, the metal expands and so does the hole.

Picture it. First put the wagon wheel on its axle. Heat the washer to expand the hole. When the hole is wide enough, slide the washer onto the axle. As the washer cools, it contracts snugly onto the axle. No need to run to a hardware store, which were few and far between for pioneers.

APART FROM BLOWING off Ajay's clothes, rapid heat expansion can crack cups. Tea cups are made of ceramic; so are coffee cups. Tea cups are thin; coffee mugs, thick. Let's borrow Ajay's thick ceramic mug, the one that says "Hot Stuff." Heat moves slowing through ceramic. That's another way of saying ceramic is a poor conductor of heat.

Pouring hot water into "Hot Stuff" is again like lightning. The inside expands rapidly, but the mug conducts poorly and so is slow to warm. That creates stress between the hot interior and still cool exterior. Under stress, the mug will crack. Under stress, many crack.

Rather than cracking Ajay's favourite mug, we could use a tea cup. Thinner cups reduce the temperature difference, inside to outside. There is less ceramic in the cup wall, so less difference. Another options is to put a metal spoon in "Hot Stuff" before the hot water. The metal absorbs some of the heat, so there is less to cause stress.

Under the tree, Ajay might wonder whether there is inverse lightning. Would quickly cooling the inside of "Hot Stuff" crack it from the outside-in? Dry ice is almost a hundred degrees below zero. A cool idea to think about on a hot summer's day.

CHAPTER 2
COLD MAKES WATER EXPAND

Earth Tilt, Pop cans, and 7th Inning Breeze

WINTERS ARE cold in the Northern Hemisphere. Jaya, next-door, says that's because Earth is farther from the sun.

JAYA

Earth wanders out even farther some years, so those years it is colder than usual.

Ajay doesn't see it that way. Sure, the Earth orbits old Sol. Not in a circle; more of an ellipse, like a watermelon. But Earth does not stand up straight to face the sun. It tilts a bit. It leans back from the sun in winter. Tilting away makes the sun lower in the sky.

AJAY

Lower sun makes for shorter days, less heat from the sun. Less heat from the sun is the cold of winter.

Cold is the absence of heat, much as a shadow is the absence of light. Cold does not exist on its own; it's a measure of what is missing. Less heat, more cold. So, Earth tilt makes winter cold.

~

As they cool, most objects contract. Metal cans, for instance. Water and rubber are exceptions. They expand when chilled. Sometimes explosively.

AJAY

Grape windows, yes. I remember. Last winter, I put a couple cans of grape pop on the ledge outside to cool. I forgot about them and, in the morning, found a spray of frozen grape soda all over the window.

Pop is water. Water expands. It was expanding as the can was contracting. Boom! Grape windows. That's bad news for Ajay, who had to clean it up, but good news for fish. Good because ponds and lakes freeze from the surface first. If they froze bottom up, ice build-up would leave little room for fish.

As water cools, it expands, spreads out. More volume, less density, like a chunk of styrofoam. Big, but light. Cooler, less dense water rises to the surface. Warmer, more dense water sinks.

As colder, less dense water rises to the surface, the surface water reaches the freezing point first. So, the pond or lake freezes from the surface down. At the same time, the relatively warmer water pushed down keeps the fish from freezing.

Cold on top gave Ajay an idea on a summer fishing trip. He put cans of soda pop in a tub that has no lid.

I bought a block of ice at the bait and tackle shoppe. To keep the cans cool, I put the block on top.

Warm air around the cans rises. Rising up, it bumps into the ice, cools, shrinks, then sinks due to gravity and the cycle repeats. This is a convection current, circulating warmer and cooler fluids.

~

The fishing trip was successful. After dinner, Ajay built a fire in the cabin fireplace and tuned in a baseball game on his radio.

The windsock showed no breeze outside. Even so, around the 7th inning stretch, I could feel cool air enter by the open window.

Ajay, the fireplace, even the radio give off heat. Warm air in the cabin rises and exits out vents in the roof. Cooler evening air from outside moves in to fill the gap. As evening draws on, it becomes cooler outside than inside.

Something similar occurs outside by the lake, often called sea breeze or lake effect. Land warms faster than water, so warm air above the land rises as cooler air above the lake moves in.

On shore, I feel cool air from the lake. Lake effect. In my cabin, I feel cool air from outside. I call this the 7th Inning Effect.

CHAPTER 3
COLD IS THE ABSENCE OF HEAT

Fridge Doors, Root Cellars, and Heat Leaks

ARRIVING at his cabin on the outset of winter, Ajay hears a hum and finds it warmer inside than expected. His nephew was here last weekend and must have left the refrigerator door open. The motor is straining.

A refrigerator cools, but it doesn't work like an air conditioner. The motor gives off more heat than is absorbed by refrigerated air. If left unplugged and the door open, that would cool off the cabin, but only for a while.

Ajay restocks the 'fridge and heads to the cellar, where he keeps a stock of potatoes, onions, garlic, beets, carrots. Most he grew in the autumn. Freezing temperatures are expected overnight.

AJAY

I keep a big pot of water near my vegetable bin to protect the produce from freezing. I borrowed this bit of wisdom from the shed, where I keep my car ready to start even in winter. The shed doesn't have a block heater for the car.

A block heater plugs in and a trickle current keeps the radiator warm enough to prevent freezing. Instead, Ajay keeps a tub of water near the radiator. Water has a certain amount of heat, keeping it liquid. The water might not feel warm, but hidden (latent) heat is there. Water releases latent heat as it turns to ice, so the tub of water is a heat reservoir.

BEFORE LEAVING, Ajay's nephew turned down the heat in the cabin. Not off, but he did set the thermostat low. Jaya, next door, says it takes more energy to reheat the house when you return, but that's not how Ajay sees it.

AJAY

The cabin is like a leaky bucket that you have to keep filling so the water stays a certain level. Due to leakage, less water is needed to refill an empty bucket than to constantly keep it at the same level.

Likewise, less heat is needed to reheat a house than to keep it warmer than the out-of-doors. Fancy analogy, says Jaya. Simple idea, Ajay replies. Any house loses heat when it is colder outside than inside. A house continually loses heat, so it has to be continually reheated.

JAYA

Maybe so, but turn down the thermostat and you let in the cold. It comes in like a wave just opening the door. I feel a chilly draft now.

AJAY

It isn't a wave. You don't let it in. Cold is an absence, like a shadow or donut hole.

JAYA

My fridge isn't cold, just not warm? Then what keeps outer space 100s of degrees below zero?

Heat is a measure of the motion of molecules. In a total vacuum there are no molecules to vibrate, so temperature becomes meaningless (and so does time, which is a measure of relative motion).

Where there are molecules like radiation left over from the Big Bang, cold is still the absence of heat. At absolute zero, there is no heat, no motion.

AJAY

Some winters, it feels like absolute zero in my cabin.

CHAPTER 4

CONDUCTION CAN FEEL COOL

Double Blankets, Chilly Tubs, and Baking Tins

HOT AIR RISES. That's how a hot-air balloon works. As hot air rises in a chimney, it creates a draft (pulls in air) at the base. That sucks air out of the room. This can cool a building even on hot days with no wind.

During the day, the sun heats the chimney, especially if the chimney is painted black to absorb heat. Warm air inside rises up the chimney, creating an updraft at the base that pulls cooler air into the room or building.

Ajay's cabin has a loft, but Alan and Lana like to sleep in front of the fireplace. When the fire goes out, so goes the heat radiated by the fire. However, the draft continues to draw warm air out of the room.

With warm air pulled out of the cabin, blankets are needed. The cabin has thick and thin blankets. A thick blanket is an efficient heat insulator, which is to say it resists heat passing through. A thin blanket is a poor heat insulator; it lets heat through easily. Ajay has a couple blankets for each person: one thick, one thin, and in various materials.

Alan's blankets are both cotton and he put the thick blanket on top, figuring it will block the heat from rising. He slept warmly. Since his blankets are made of the same material, it does not matter which is on top. The blankets are heat insulators in series. That is, the heat must go through both before it escapes.

Lana has a thin satin blanket and a thick cotton blanket. She put the satin blanket next to her. Lana felt chilly, even though the temperature under the covers was the same regardless which blanket is on top. She felt cooler with skin touching the satin than if touching the cotton. Satin conducts heat away from her, more than the cotton would have. Same temperature as cotton, but satin is a better heat conductor.

In the morning, Lana went for a bath. Leaning on the tub to turn the tap made her feel chilly. The wood door, bath mat, and towels did not feel cold, but the tub did even though all are in the same room. The tub is cast iron coated with porcelain enamel. Metal conducts heat away, so the tub feels cooler, even at the same room temperature. The tub felt like it was lower temperature—like the satin blanket.

ALEX IS BAKING muffins and tests whether they are done by inserting a toothpick. If the toothpick comes out clean, the muffins are done. He reaches into the hot oven barehanded and presses in a toothpick. The oven is several hundred degrees, but he isn't burned when reaching in, testing the muffins.

With the last batch, I touched a thumb to the muffin tin.
A painful blister popped up almost immediately.

The air in the oven is the same temperature as the muffin tin, but air has very little mass. The atoms are more spread out, not many atoms bumping against one another, so air is a poor conductor of heat.

The metal tray has a lot of atoms packed together, bumping into their neighbours. As a result, it is thousands of times better than air at conducting heat. Air is an insulator; metal is a conductor.

After the blanket and tub experiences, Lana was happy to have something warm in hand. She held the fresh muffin, letting its heat conduct into chilly fingers, inhaling sweet convection currents.

CHAPTER 5
HEAT FLOW CAN BE SLOWED

Cast-Iron Cakes, Hot Coffee, and Fresh Snow

Lana found a cast-iron skillet in the cabin cupboard and slid it over a stove element, preparing to make pancakes. After a while, she plopped a drop of water into the skillet to test its temperature.

The drop sizzled and hopped around the hot pan. In a hot skillet, the bottom of the drop turns to a vapour cushion. The drop rides above the pan, much as a puck hovers over an air hockey table. When the water drop dances, Lana knows the skillet is ready for pancakes.

LANA

This test won't work with a drop of vegetable oil. Oil can't form for a vapour cushion and would flop onto the skillet, however hot.

Lana likes thick, cast-iron skillets. They have a uniform temperature across its bottom. Many pans are thin steel. While lighter, a thin pan can have hot spots over the burner, making food stick to the pan.

ALEX POURS his uncle a cup of coffee.

> AJAY
>
> Thanks. I will be back in a moment. I need to fetch fire-wood, but go ahead and add the cream.

> ALEX
>
> I'll wait. The cream is cool. If I add it later, the coffee will stay hotter.

Not so. To keep the coffee hotter, add the cream now. That which has more heat, has more to lose and at a faster rate. The coffee and cream combination loses heat at a slower rate than hot coffee alone.

In any case, Ajay won't have to wait. Alan and Lana offer to fetch kindling and an armful of logs from the shed. The weather has been dry and cold for days, resulting in static sparks putting on clothing.

Alan and Lana notice how quiet it is outside. Last night it snowed. Fresh snow is not packed. The fluffy stuff has many small spaces between flakes. Air spaces trap sound waves like acoustic tile. Even so, their feet make a crunch-crunch sound across the snow. If a bit warmer, pressure from their weight would partially melt flakes under foot and lessen vibrations, so there would be no crunch.

Below -15 °C is too cold for snow crystals to melt under pressure. Instead, snow crystals slide past one another, compressing with a crunch. When it is too cold to melt with weight, it can be difficult to ice skate. Lana mentions the snow on return.

That's good news. We have one last crop to bring in. In this cold, rain would freeze on contact.

Freezing cell fluids burst their walls, ruining the crop. Fresh snow is protective, like an igloo. Trapped air pockets are a poor conductor of heat, so a snow blanket insulates the ground's heat, protecting the crop from much colder air above the snow.

CHAPTER 6
LIQUIDS CAN REDUCE FRICTION

Tire Traction, Stone Cold, and Sand Sliding

AJAY AND ALEX took the truck into town to pick up supplies. Before heading out, they noticed the tires looked a little flat.

Most matter expands when heated, pushing out against a container. That is why basketballs bounce higher on hot days and are flatter in the cold.

Most matter shrinks when cooled, which why the snow tires on the Alex's truck appear a bit flatter today. Flatter, they have more contact with the road, better traction.

Good tires grip by creating friction. Liquids, less so. That which flows has less friction than solids. Oil, for instance, can make a surface slick. So can rain. That's important since ice can't take much pressure. It melts and the melted layer is slick.

Snow melts unless it is what Ajay calls stone-cold: too cold to pack a snowball. Packing requires pressure melting, at least on the surface, to bond the snowball together. That's not possible when stone-cold.

When it is stone-cold outside, ice is like pavement, making it is too cold to skate.

It would be like trying to skate on dry ice (solid carbon dioxide). Lots of grip, but little glide. Skates need to glide over a thin liquid layer, but dry ice does not melt into a liquid. It sublimates from solid to gas. When not so cold, standing on ice pressure melts to a thin layer of liquid, slick as oil.

So walking on ice is walking on water, without the sinking part. That's why ice can turn a sidewalk into a slide-walk. It's slippery when wet. Walking on bumpy ice is even more difficult with little grip by friction.

Bumpy ice has with fewer points of contact; mostly at the top of the bumps. Less contact can be helpful with ball bearings or over-filled bicycle tires. Less contact means less resistance due to friction.

Walking on the moon is more difficult than walking on Earth because of less friction between shoes and the surface. Weight on the moon is one-sixth of the weight on Earth. That means one-sixth of the grip.

Walking in a backyard pool is more difficult than walking on the deck. Water in the pool buoys up a person. Being pushed up causing less traction between a person's feet and the bottom of the pool.

Walking on beach sand is more difficult than walking on a boardwalk. On a beach is more difficult because sand grains are not connected and act like a fluid. They provide little push-back (traction).

Little traction or not, in the stone-cold weather of mid-winter, Alex and Lexa wish they were on a warm beach somewhere. Alan and Lena, however, want only a few degrees warmer to glide across the pond.

CHAPTER 7

INSULATION REDUCES HEAT FLOW

Silver Warmth, Ceiling Fans, and Walking on Fire

LANA SWITCHED to cotton after her experience with the satin blanket. She also puts a hot water bottle under the covers on chilly nights. The bottle will stay warm for hours since water resists heat changes.

That water resists heat changes is also why the water drop danced in Lexa's skillet. A drop of oil would pass heat through itself quickly compared to water. Instead, the water heated and evaporated on one side.

Cool water likewise resists warming. Food that has a lot of water, such as melon, will stay cooler for longer than other items in a picnic basket. Foods with oils, such as potato salad or coleslaw, spoil faster.

Compared to metal, water needs a lot of energy to warm up. Water also gives off a lot of heat as it cools. Ten times as much as iron. Twenty times as much as silver.

Some things we want to warm quickly. Silverware, for instance. Silver is desirable for cutlery since it warms quickly in a person's mouth. Gold warms even faster, twice as fast as silver.

Alex and Lexa had wine with supper, served in long-stem wine glasses. They keep wine refrigerated and say it tastes better cooler. A wine glass has a long stem to prevent a person's hand warming the wine.

Ajay filled his "Hot Stuff" mug half with water, half ice cubes. When the cubes melt, there won't be a spill. The cubes push aside an amount of water equal to their weight. As they melt, the level of liquid stays the same.

If the ice cubes contain a lot of air bubbles, as often happens with tap water, there will still be no difference on melting. Heavier or lighter, they push aside (displace) as much water as is equal to their weight.

The great room of the cabin has a large fireplace and an equally large ceiling fan at the apex of the pitched ceiling. To set the air draw for the fan, up or down, Ajay pulls a cord dangling from the fan.

Ceiling fans do not change the temperature in a room. They circulate air. To feel cooler, set the ceiling fan so it will pull air upward, drawing warm air up away and so make a person feel cooler.

Warm air from the fireplace rises, bringing cooler air in and onto those in the cabin. To feel warmer, Ajay pulls the cord on the ceiling fan so it pushes the rising warm air back downward.

∽

Warm air rising is one reason the insulation is typically thicker in a roof or attic than in the walls and why there is little or no insulation under the flooring.

Insulation, whether it is a foam or a blanket of fibre, contains many tiny air pockets. Air is a good insulator because its molecules are spread out and so do not conduct heat.

A goose down blanket or down coat has air pockets and so retains warmth. Even crumpled paper stuffed into coat sleeves will hold in a person's body warmth better than the sleeves alone.

A blanket or insulation or igloo don't produce heat. They reduce heat loss. A blanket around a thermometer will not raise the temperature since neither blanket nor thermometer produce heat.

AJAY

Insulation is the secret to walking over hot coals. Wait until they burn down to have a layer of ash. Ash, like fresh fallen snow, has pockets of air which resist passing heat along. But walk quickly.

Walking over hot coals is easier if the coals are from wood. Wood is a poor conductor of heat. Nervous sweat may also help as water resists heat change. Lana figures she will stick with the hot water bottle.

CHAPTER 8
CONDUCTION IS HEATING BY TOUCHING

Efficient Engines, Icy Bridges, and Popsicle Sticks

THE WINTER FAIR has an auto show. One company at the show has on display what they claim is a 100% efficient motor.

ALAN

Can't be. Just feel it. A completely efficient motor would lose no heat.

ALEX

Noisy too. An engine that is 100% efficient would lose no energy to heat or through vibration.

Ajay, catching up with them, watched as a technician added fuel to the engine.

AJAY

An all-efficient engine would not need fuel to run or oil to reduce friction, nor would it waste any fuel as exhaust. It would also be a perpetual motion machine.

~

THEY STEP OUTSIDE into the freezing night air, each wearing a different kind of coat. Ajay's puffy coat traps heat. Air is a poor conductor, so Ajay's body warmth isn't lost to the night air.

Alan's leather jacket looks cool, but makes him feel cold. It offers little protection: a single layer, no lining or padding, and large conductive metal zipper. A flannel shirt under keeps him from freezing.

Alex's coat has a thin layer of insulation to reduce heat loss. It also has a reflective material sewn inside to radiate body heat back to Alex. The reflective material is similar to the shiny side of aluminum foil.

~

OUT IN THE NIGHT AIR, Ajay, Alan, and Alex can see vehicles for the winter fair in various states of preparation. One car has no wheels and the chassis sits flat on the ground. This one has no frost on it.

Heat from the ground rises into vehicles touching it, such as the one with no wheels. But rubber tires prevent heat from being conducted from the ground and so the cars on tires are frosty.

ALEX

Driving back to the cabin, let's avoid the bridge. Roads can be icy over bridges.

Earth radiates heat, but there is no ground below a bridge to conduct heat up to the pavement and prevent ice. It's like mittens. Mittens keep fingers warmer than gloves on cold days since, in mittens, fingers touch one another and so share heat. Or like snuggling two in a sleeping bag or more in a tent on a frosty night.

On a frosty night, frost forms on the tips of grass. It doesn't form on dirt, stone, or other heat conductors. Heat from the ground conducts into stones and trees, so the frost is less likely to form around them.

ALEX

Heat can be conducted quickly, having learned the hard way as the kid who licked a flagpole in freezing weather.

ALAN

Lick a popsicle and your tongue might also become stuck.

AJAY

Heat is conducted from tongue to the popsicle or flagpole. Stick with the popsicle stick. It won't stick since wood is a poor conductor of heat. Kind of like walking on ash over hot coals.

CHAPTER 9

COMPRESSION WARMS; EXPANSION COOLS

Hot Springs, Sliding Doors, and Boiling Spaghetti

LANA AND LEXA went to the mountain-top spa, part of the winter fair. Warm air rises, yet it is colder on the top of a mountain than at the bottom. High up, air loses heat to outer space faster than it can be warmed by the sun.

But that's not the whole story. As air rises, there is less air pressing down from above, so it expands. With particles moving further apart, they have fewer collisions causing heat and so the gas cools.

Ascending the mountain is like releasing air from the tire. Pumping up a bicycle tire pressure heats the air inside the tire. Releasing air spews cool air out the value. Cool because it is rapidly expanding.

Lana figured out which way the breeze is coming from by holding a wet finger in the air. Evaporation is greater on the windy side, which will feel cool. The cool side of Lana's chilly digit is toward the breeze.

~

STEPPING INTO THE WARM SPA, Lexa's glasses fog up. Heat passes into the lens, leaving moisture to condense on the surface of her lenses. Fogging from warm breath happens when she wears a surgical mask.

It is hot in the cedar spa and both of them sweat. Sweat evaporates and carries heat away so a person feels cooler. This sulphur spring is 37 °C, the same as a person's normal internal temperature.

People perspire, but there is no evaporation in the water, no cooling. To avoid having customers become ill from over-heating, the spa limits how long patrons can be in the sulphur spring.

LANA

Walking to the restaurant, I saw a cloudy sliding glass door. It must have a crack.

Double pane windows and doors have a layer of nitrogen gas between the panes to prevent heat from passing through. A crack lets in air. Air contains moisture, which condenses and causes the panes to become cloudy. The process is similar to Lexa's foggy glasses. Warmth passes through glass, leaving behind moisture.

If there is warm inside and cold outside, the warmth passes inside to outside, leaving the moisture indoors. If there is warm outside and cold inside, the warmth passes from outside to inside, leaving the moisture outdoors.

LANA HAS soup and salad for lunch. The soup is too hot. Blowing across the top of the soup increases evaporation and removes the warm vapour layer that works like a lid to hold heat in the soup.

Lexa ordered spaghetti. Spaghetti is often cooked in vigorously boiling water so the spaghetti strands don't stick to the pot or to one another. A gentle boil is better for cooking macaroni or soup, or potatoes.

Whether gentle or vigorous, water boils at the same temperature: 100 °C. Anything more than that is carried away with the steam. Vigorous boiling does not cook spaghetti better or faster.

LEXA

At home, I boil spaghetti gently, stirring from time to time. Once the water starts to boil, I will turn down the temperature to a gentle boil.

Stirring takes the place of a vigorous boil, but a pressure cooker is even more efficient. It cooks at 120 °C by preventing steam from escaping. The boiling point of water goes up as pressure increases, so it cooks faster and uses less energy in the process than conventional boiling.

PART 2: SPRING INTO ACTION

Matter and Motion

CHAPTER 10
OBJECTS ARE LAZY BY NATURE

Lazy Tomatoes, Stable Wheels, and an Old Magic Trick

WHILE OUT GROCERY SHOPPING, Ajay pulls on a large roll of plastic bags near the vegetables. The big roll at rest resists unrolling.

AJAY

When I pull slowly, the rolls unwinds. The bag stretches out and tears off unevenly.

Next time Ajay uses a quick snap-jerk. Snapped, the bag comes off the roll cleanly. A snap-jerk applies force along perforation holes between bags on the roll.

The vibration causes a couple tomatoes to roll off the table together. They hit the floor at the time. The heavier tomato is twice as heavy. It is also twice as lazy, which is to say it has double the inertia of the smaller tomato.

If tossed off a diving board, likely a tomato and I would splat at the same moment.

He is right about that. More mass, more pull of gravity, but also more inertia. Gravity and inertia even out.

THE STEEL SIDE door of Ajay's house closes by a spring. He pulls it open to enter. As he goes through, arms loaded with grocery bags, the door swings back and hits one hand. That doesn't hurt, at least not when carrying paper towel or anything light.

Carrying a crate of potatoes, however, the door is painful as it strikes the back of his hand. The potato crate is more massive, so there is more force in the impact. More mass, more force.

ALAN AND LANA rode over to help prepare the meal. Their bicycle wheels help keep their bicycle stable since they resist tipping. They are lazy that way. Each wheel is a large gyroscope, like a spinning top.

Karl's motorcycle is even more stable due to faster spin of the wheels. Simplified, a spinning wheel acts to keep spinning and resists tipping. The faster the spin, the greater the resistance to tipping.

ALAN SET the table using a silk table cloth with a checkered pattern. Glass rubbed against silk becomes positively charged and can create a static spark, but Alan has something else in mind for after dinner.

The potatoes will be boiled, but first must be washed, then cut. Lana opens the tap. As the stream of water falls from the faucet it becomes narrower and breaks into droplets just above the basin of the sink.

Water picks up speed as it falls, going further with each second. The same amount of water comes from the faucet each moment, so the water stretches, like stretching chewing gum becomes thinner.

LANA

Cutting potatoes into cubes works better with a cleaver than with a knife.

Both are sharp, but the cleaver is more massive and so it strikes with greater force. More mass, more force. It takes more effort to swing the cleaver, but once in motion it is easier to keep the clever in motion, splitting potatoes. The clever is lazy. Staying at rest or staying in motion.

Lana finishes by slicing the cubes so there is more surface area exposed to the boiling water. More cooking surface means less cooking time. It is like splitting wood into kindling for more surface area.

After dinner, Alan snaps the table cloth out from under dishes. The dishes are lazy, much like the roll of plastic bags back at the grocery. They stay put as the tablecloth is pulled out from under them.

It helps to pull the cloth out and slightly down. If pulled upward, the cloth would tip over upright glasses.

It also helps that the cloth is slick and seamless and the glasses weighted down with water. More mass, more laziness.

CHAPTER 11
OPPOSING CONTACT CREATES FRICTION

Wrestlers, Yo-Yos, and Shooting Stars

ALAN IS HEADED to the gym with Clay and Lacy. The siblings consider themselves retro. Alan considers them nerdy. A skateboard made from actual skates. A bicycle with banana seat and butterfly handlebars.

Retro or not, Clay isn't likely to slide off his skateboard. With the size of his foot, his sneakers make a lot of contact with the board. More contact means more friction, foot to board. More friction, less sliding.

The skateboard wheels have less contact with the sidewalk than a sneaker has. Much less. Each wheel touches the sidewalk along a thin line segment of contact. With little friction, wheel to ground, the wheels roll easily.

Clay has his skateboard, Alan and Lacy have bicycles. Hers has white-wall tires. Lacy over-inflates her tires for less contact with the road, like Clay's steel skateboard wheels. Less contact means easier to roll, easier to skid.

CLAY

> I have an idea for a skateboard that uses big ball bearings instead of wheels.

Spheres have even less surface contact than wheels. Less contact, less friction. His plan has three ball bearings. Two in front; one at the back. That gives stability with minimum contact.

LACY IS TALENTED in gymnastics and yo-yo tricks. Alan and Clay call her Morpheus, after the Greek god of dreams, since she is a master of making a yo-yo sleep. Modern yo-yos, like the one Lacy uses, have ball bearing to reduce friction between the string and axle.

A basic yo-yo is two discs connected by a peg (axle) and string tied at one end around the peg. A yo-yo spins ("sleeps") when there is little friction between the string and bottom of the peg—not enough for the yo-yo to wrap around the string.

Jerking the string up forces it to make more contact with the peg. The string sticks to the peg by friction and wraps around it. As it wraps, the string becomes shorter and up goes the yo-yo.

ALAN, a wrestler on the school team, begins by rubbing his hands together, warming them up by friction. His hands resist sliding, producing heat.

Same thing happens when a tire hits the road. The molecules on one surface form links of attraction with another surface and resist sliding. Friction comes from the links between the tire and asphalt. Pulling them apart gives off heat, so tires that have travelled for a while are warm.

When rubbing his hands together, Alan tries to push harder on one hand than the other. The exercise never works. Each hand pushes equally on the other. The action on one is an equal action on the other.

Once warm, Alan applies powder to keep his hands dry. Wrestling is sweaty business and it is best to keep a grip on the opponent. Alan uses friction to grab his opponent, but lack of friction as a strategy.

Alan's wrestling strategy is to get under his opponent and push upward, lifting his opponent off the floor. Lifting means the opponent doesn't have ground to stand on and no friction to resist.

~

AFTER THE MATCH, outside the gym, the trio spot a shooting star. It isn't a star; more likely a grain of sand from outer space that burns up in the atmosphere by friction against the air. Bright, but brief.

As they take off for home, the two on their bikes pull up on the handlebars. The handlebars, in turn, pull down on them and this force is transmitted to the pedals for a quick take-off. Home before dark.

CHAPTER 12

MORE TIME, LESS IMPACT

Gym Mats, Catcher's Mitts, and Hammer Hard

Lacy's gymnastic practice takes place on a basketball court. The court has hardwood floors, but a springy layer underneath the hardwood. This layer spreads the impact over a longer time, reducing the force.

The gymnasts have mats to lessen the impact. Impact spread over a longer time reduces the force, so less chance of injury. The gymnasts also bend their knees to lessen the impact, like compressing a spring in a mattress.

Padding to extend the time of impact is commonplace: boxing gloves; rubber bumpers; carpeting; pillows; automobile airbags, padded dashboards. It is even in automobile hoods that crumple on impact.

Some apartments, especially in older buildings, have an unwanted example of springy floors. The thud and thump sounds from the apartment above are made by people walking, causing the floor to vibrate like a drum.

Newer apartment buildings have a slab of concrete between the floor of one apartment and the ceiling of the apartment below. The slab has no spring to it and so it prevents the thud and thump effect.

～

WHEN WRESTLING, Alan will roll with a lunge from his opponent. That's what a boxer does: roll with the punch. Rolling increases the time of impact, which means less force in the punch.

Increasing time of impact to reduce the force is common in sports to prevent injury. Baseball catcher's mitts, for instance, have more padding than the mitts of other players. Padding prolongs the time of the impulse to stop the baseball and lessen the force on the catcher's hand.

Mountain climbers will often use nylon ropes since they stretch. Stretching increases the time to bring a falling climber to a halt. Less force means the rope is less likely to snap.

Bungee cords for jumping stretch a lot to reduce the time of impact. A steel bungee cord would likely snap. The stop time is too abrupt. Similarly, a concrete floor is more tiring than a wood floor.

Modern automobiles are made to crumple on impact, extending the time of impact, absorbing some force of the impact. Race cars are made to blow apart and so carry away the force of a crash.

As mentioned, an automobile hood that crumples can prevent serious injury to a pedestrian struck by the car. Likewise, roadway guard rails are designed to bend. Light poles on a highway are surrounded by barrels which crush to slow the impact.

I can throw an egg against a bedsheet hanging on a clothesline and it will not break.

But throw it against a wall and splat! An egg that falls on carpet is less likely to break than if it falls onto a hardwood floor.

The more massive an object is, the greater the force it has when it hits. A cruise ship might not be fast, but it has a lot of mass. To dock safely, the engines stopped still far from dock.

Sometimes it helps to have more impact over less time, like hammering in a nail. Swinging a hammer gives it kinetic energy, so it strikes the nail with a blunt force, driving the nail into the wood.

It is not possible to push in a nail and my hand is a poor and painful substitute for a hammer.

Skin is soft and cushions impact, unlike a hammer. The nail goes into Clay's thumb rather than the wood.

CHAPTER 13
SOLIDS HAVE STABILITY

Finding Balance, Slicing, and Rope Walking

THE WOLF PACK is an outdoor adventure group. Members do a lot of hiking, camping, and woodcraft. Often their gear laden backpacks are heavy and awkward to carry. Noam, Lacy's dad, is their pack leader.

NOAM

Lean into it. Bend forward when carrying a heavy load on your back.

Leaning forward shifts the centre of gravity of the load to above the feet. If the weight were shifted backward, a person would fall backward.

They set up camp on the edge of a forest, not far from a stream. Everyone is assigned chores, one of which is to fetch several buckets of water.

Carry two buckets at a time, not just one. It is easier that way.

It is easier to carry the same amount of water in two buckets, one in each hand, than in a single bucket. A person can stand upright since the centre of gravity is between the feet so one does not have to lean.

To start a campfire, Noam tells them to think thin. Kindling, not logs of wood. Kindling has more surface area for the same amount of mass. It will reach a higher temperature in less time than the logs.

Thin applies to supper as well. Potato slices cook faster than wedges. Flatter burgers cook faster than thicker burgers with the same mass. Thinner and flatter expose more surface area for cooking.

While out picking mushrooms, some members see small animals curled up. Think thick, says Noam with a smile. Animals curl up into a ball since curling up exposes less surface area to the surroundings.

The troop made caramel apples. It takes less candy to make them using larger apples, one per member, than using smaller apples, two per member, since larger apples have less surface area per weight.

MORNING STRETCHES INCLUDE SIT-UPS. Sit-ups with knees bent is easier than sit-ups with legs straight out. Straight out, your centre of gravity is farther away, so you exert more torque sitting up. Torque exerted at the base of the spine is more prone to injury. After stretches, the troop walk a tightrope tied between two trees.

NOAM

Shake a stick at it. By stick, I mean a pole.

A pole long enough that it droops at the ends lowers the centre of mass of the tightrope walker. With a drooping pole, centre of mass may be below the tightrope. The pole also resists rotating, keeping the tightrope walker stable on the rope.

The tightrope is twice as thick as ordinary rope. Twice as thick has four times the cross-sectional area, so is twice as wide and four times as strong. With the rope and pole, all members are successful.

CHAPTER 14
FASTER FLUID, LOWER PRESSURE

Rocking, Rising, and Measure of Pressure

CLAY'S MOM, Mona, volunteers on the school bus when there is a gymnastics competition. She sits near the middle. On a bumpy ride, the front and back of the bus rock up and down more than the middle.

Rocking rotates the bus about its centre of mass, which is just in front of the middle. The engine weighs down the front, so the back of the bus is lively. Some kids choose the back for this reason. On a bumpy road the bus is like a ship in a choppy sea or airplane in turbulent air.

MONA

The farther I sit from the centre, the greater the see-saw up and down motion.

~

Other times no bus is available and Mona drives Lacy with a few other gymnasts to the competition. As they accelerate to highway speeds, she asks the girls to roll up the windows.

Open windows cause drag. Running the car's air conditioner increases fuel consumption, but at highway speeds, open windows create air drag that uses more fuel. Drag could offset any saving from turning off the air conditioner.

On the highway, Mona's car slightly leans toward a passing truck. Air between them is moving fast and the faster a fluid flows, the lower its pressure. With lower pressure between the car and truck, surrounding pressure pushes the vehicles toward each other.

Wind across the top of the chimney at Ajay's cabin likewise lowers pressure inside the chimney, creating a better draw from air inside the room. That helps the fire burn, but also draws out warm air from the room.

It wasn't until they stopped for lunch that Mona made the connection. Lacy, playing with a straw in a glass of lemonade, blew across the top of the straw. The fast air drew lemonade up the straw.

That's why prairie dogs make mounds around their entrance and exit holes. As air is forced up and over the mound, air pressure over the hole lowers, pulling up air from the tunnel, providing prairie dogs with ventilation.

❧

After lunch, Mona still has ice in her lemonade. The ice in Lacy's drink has melted away. That's because Mona had ice cubes and her daughter had crushed ice in their lemonade drinks.

Crushed ice melts faster than ice cubes since the surface area of crushed ice is greater, which gives more melting surface to the surroundings. It is the same principle as kindling or potato slices.

The glasses were poured to the brim, but neither over-flowed when the ice melted. When ice be melts, the water level is unchanged. The volume of water displaced by the weight of the cubes is the same.

Cloudy ice cubes make a slight sizzling sound as they melt. They release the small pockets of air that make the cubes cloudy. Clear ice cubes can make a cracking sound when placed in liquid.

Cracking occurs by a sudden change in temperature. Water is a poor conductor, so the surface of the cube and its cooler interior are under tension. The cube cracks, similar to Ajay's "Hot Stuff" mug.

A WOMAN ENTERS the restaurant and takes off her high heels walking across the restaurant's wood floor. If worn, her high heels could ruin the floor. Their pressure is in a smaller area than an elephant's flat foot and so can have more pressure.

When the door opened and the woman entered, the girls caught a whiff of gasoline. In the lot, a car is leaking fuel from its gas cap. A tank filled to the brim can overflow if the car is then parked in direct sun. The gas expands with heat, bursting through the gas cap.

CHAPTER 15
SMART BODIES SELF-REGULATE

Smart Home, Sweat, and Stinky Sneakers

LACY LIKES GYMNASTICS. Clay plays basketball. Both like coming home to the comfort of a smart home. Lights automatically on at dusk, off at dawn. They can give a voice command for a favourite playlist or open the garage door.

Heating and cooling are part of the smart system. The thermostat has a strip that is brass on one side, iron on the other. It bends as it gets warmer, breaking the circuit which kept the heater on. As the strip cools, it bends back. When it makes contact, completing an electrical circuit, then the heater switches back on. The dehumidifier like that, but with a belt instead of a bimetallic strip.

In a dehumidifier, a fan pushes air over a nylon belt. Higher humidity causes the belt to stretch and trip the ON switch. Lower humidity causes the belt to shrink and break off the ON connection. By use of feedback, the thermostat and dehumidifier keep the temperature and humidity inside the house within a set, comfortable range. The system isn't smart as much a responsive to changes.

∿

CLAY AND LACY are healthy and maintain a stable internal temperature of 37 °C. Doing work causes food to be converted into motion and heat. Excess heat can be unhealthy, so they sweat to carry away heat.

Sweating works for people, but is not enough for a bulky animal because it has a large mass relative to a small surface area for cooling. That's why warm-blooded animals are limited in size and dinosaurs were much larger.

An elephant has a lot of mass and relatively little surface area. Living in a hot climate, it needs a large surface area to dispose of extra body heat through radiation and evaporation. Large ears are large radiators.

Every day, Clay and Lacy eat about 2% of their own weight in food. The mouse in the attic eats 50% its own weight in a day. In other words, 25 times as much as people. Even so, there are no obese mice.

Compared to an elephant, a mouse as little mass and a lot of surface area. The elephant needs to lose heat. The mouse needs to make heat, so it requires extra food energy to maintain body temperature.

The heat loss in animals is mainly through their surface, so it varies in proportion to an animal's surface area. Heat production, however, occurs in all cells and varies in proportion to an animal's volume.

Babies have a larger surface area for their mass than adults do and so dissipate more heat. They feel colder at the same room temperature. Layers of bundling help insulate against excessive heat loss.

More surface, more heat loss. More heat loss, then proportionally more food is needed to maintain healthy body temperature. That's why a mouse needs proportionally more food than an elephant.

~

CONTROLLING fluids in a living body is another type of self-regulation. The entire blood supply is filtered by the kidneys 60 times each day and about 99% of the water filtered is reclaimed by the body.

In human beings, the lungs, skin, and urinary system work to expel wastes produced in metabolic activities. Lungs expel water and carbon dioxide. Sweat glands expel water, salts, and urea.

CLAY

> I wish my basketball shoes were self-regulating. They stink.

Water, salts, and urea from the skin can become trapped, such as in an arm pit or inside a shoe. An accumulation of urea causes an unpleasant odour.

Walking barefoot or not washing causes an accumulation of bacteria, which contribute their own waste odours. Playing ball, Clay loses as much sweat as the cooling equivalent of a small air conditioner.

The body has more than two million sweat glands. For a healthy person, sweat is mostly water with little or no odour. Bacteria on the skin pollute sweat and give an odour to perspiration.

Only some apes (such as the gorilla), horses, and a few types of cows perspire for temperature control. Other animals pant. To cool off, the ostrich and emu urinate on their legs and let the liquid evaporate.

CLAY

I used to say that I sweat like a pig, but that isn't true.

Pigs have no sweat glands. They cannot cool themselves by the evaporation of perspiration. They can lose heat by conduction; that is, by contact with something cool.

To cool off, pigs cover their skin with moisture. Mud is effective in both respects and protects the skin of a pig from pests and sunburn. To sweat like a pig, clay would cool off with a roll in the mud.

CHAPTER 16

REACTION IS EQUAL AND OPPOSITE

Pond Launch, Walking through Walls, and Missing Matter

THE POND FROZE over winter and now, with spring thaw, it is super slippery. Clay has an idea for free motion. The pond is not perfectly frictionless, but is far too slick to walk on.

Alex pushes Clay from shore, launching Clay to the middle of the pond. It is too slick to walk on; even too slippery to sit up. Clay takes off an ice skate and throws it hard toward the edge of the pond. It works—he slides in the opposite direction to shore. As the skate went one way, he went an equal and opposite direction courtesy of energy he supplied. Energy from food, ultimately from the sun.

Clay has another idea. He can see through ice. He can see through glass. Sight goes through. Perhaps he can jiggle his molecules to walk through a window or a wall like a comic book superhero.

This idea doesn't work. The wall remains impenetrable. Atoms in the wall and Clay electrically repel one another. Matter might be mostly empty space, but these fields of repulsion are strong enough to make objects act solid.

Clay has a skateboard made of iron and an idea to make his iron board move with ease. Standing on the board, he holds a fishing pole with a large magnet attached where there would be bait. He hangs the magnet in front of the skateboard.

No go. The magnet pulls on the pole with the same force it pulls the skateboard. It is like trying to lift yourself off the ground by pulling up on your ankles.

Clay wants to use sand to make matter disappear. Perhaps 2%. Half a glass of water plus half a glass of alcohol does not amount to a full glass. It is missing 2%. Water and alcohol molecules fit into one another and take up less space combined than they do individually.

By analogy, if a volume of sand is mixed with an equal volume of marbles, the resulting volume will be less than the sum of their volumes. Some sand will fill spaces between the marbles.

CLAY

Water and alcohol. Sand and marbles. The sum doesn't add up to its parts. That gives me an idea of how to make matter disappear, maybe the same two percent.

He put two sand hourglasses balancing on a scale. They balance. Then he turned one hourglass upside-down.

That hourglass will be weightless during free fall, so the scale will tip to the other hourglass. What were equal weights are made unequal by motion in one hourglass.

Except he cannot do it with the hourglasses. Equal and opposite forces balance out with no net change. The two hourglasses weigh the same, in spite of the fact that some of the sand in one is in the air.

The additional force on one side is due to the impact force of the sand. This is somewhat like the ice skate thrown on the pond. In the end, Clay discovers that there is no disappearing matter and no free motion.

CHAPTER 17
TEXTURE CAN BE FABRICATED

Soft Bubbles, Crunchy Crystals, and Smooth as Cream

Occasionally, Clay will have ice cream in a bowl. Ice cream is mostly air and water.

CLAY

I stir it to make the ice cream smooth, even if it uses up some in stirring.

The stirred ice cream isn't used up, but does have a smaller volume. Stirring releases air from the ice cream. There is still the same amount of ingredients, just not as thick.

Air is put into ice cream for a variety of reasons. One reason is to make it more profitable for the vendor. Air adds volume to the product. More volume, more profit since air is free. Soft serve ice cream has even more overrun air to make it serve smoothly.

Clay thinks his stirred ice cream tastes better than when scooped out, and a lot better than premium brands that do not have much overrun. Overrun is the name given to the amount of air added to ice cream.

Whether it tastes better, it does taste more. Softer ice cream spreads across taste buds more readily than ice cream in bulk. It melts faster in the mouth, adding to detection by taste buds.

Soft ice cream is also less likely to cause brain-freeze since it melts faster. Cold food, such as from hard ice cream or a slushy ice drink, causes blood vessels at the back of the mouth to rapidly constrict.

The brain sends blood to open the vessels and conflict of constriction and expansion causes pain (similar to the stress cause by putting hot water in Ajay's "Hot Stuff" mug). Pain felt in the forehead is really in the outer covering of the brain, where the arteries at the base of the brain meet.

Cold numbs the sense receptors for sweetness, so more sugar is added to ice cream to compensate. So much so that some people find the taste of melted ice cream to be cloying, sickening sweet. The same thing is true of soda pop intended to be served cold. When chilled, the beverage tastes the right amount of sweet. Warmed, it tastes too sweet.

The smooth texture of ice cream comes from milk fats. Butter also comes from milk fats. Fat does not mix well with water (ice crystals in ice cream), so it will separate from the water as it melts. Even in frozen ice cream, fats exists as tiny globules that hold air in tiny pockets within the ice cream, like a foam. Higher fat ice cream stays on the tongue longer than ice milk and so offers more flavour.

Egg yolks keep milk fat and ice crystals together. The lecithin in eggs acts like an emulsifier so the fat globules can cluster together. Gelato uses cornstarch or tapioca starch instead of egg yolks as an emulsifier.

To keep ice crystals from forming, commercial ice cream contains stabilizers, often made from seaweed, to absorb excess water. Without stabilizers, ice cream would look like a milkshake.

CHAPTER 18
LIQUID SURFACES HAVE TENSION

Serious Grime, Surface Tension, and Walking on Water

DIRT AND GREASE together make grime, as Alex knows well, working on his bicycle. Grease is a fat. It does not come off easily by water alone. Turpentine works. Turpentine is frequently used to clean paint brushes. The solvent used for dry cleaning also works. Clothes that are dry-cleaned are washed in a container full of the chemical trichloroethane. It is said to be dry cleaning only in the sense that the tub does not contain water.

Soap also works to remove grime. Part of a soap molecule attaches to water and part to grease, surrounding a grease globule and lifting it out of clothing to be rinsed away and down the drain. Soap can be made with animal fats and caustic lye.

Detergents work like soap and are biodegradable. They break down harmlessly in the environment. There are detergents in toothpaste and dish-washing liquid. Some detergents make clothing whiter than white. A fluorescent dye turns ultraviolet light, which we cannot see, into blue light, which we can see. The clothing reflects blue, which appears as being whiter.

Soap makes suds, tiny bubbles. They are the same as the bubbles one might blow through a ring dipped into a soapy solution. A soap bubble, sudsy or blown, is a sphere of two layers of soap molecules. Air pressure makes a soap bubble round by pushing in equally on all sides. Of all shapes, a sphere has the greatest volume for its surface area.

Some extra-soapy bubbles are elongated toward the Earth due to gravity. Soap in a bubble drains by gravity, so the top thins first and is most likely to burst first. When water in a bubble evaporates, the bubble pops. Adding glycerine slows evaporation, so bubbles last longer.

Soap and detergent are less effective in hard water (water that has dissolved minerals). Clothes don't come as clean or smell as fresh if washed in hard water. Hard water also makes it difficult to blow soap bubbles.

Water is said to be hard if it has a lot of calcium and magnesium dissolved in it. These chemicals attach to the detergent molecule and reduce how well the detergent can attach to grime.

BUBBLES ARE HELD TOGETHER by surface tension. Water molecules pull other molecules from all directions. But surface water molecules have no molecules above, so they pull harder on those beside and below.

A bug known as a water strider can walk on water due to surface tension and their long, flexible legs to distribute body weight. A water strider can row across the surface at a meter per second.

Surface tension is strong enough to gently float a razor blade or paper clip. Adding a drop of dish-washing detergent lowers the surface tension and both paperclip and razor blade will sink.

Sprinkle pepper evenly on the surface of a bowl of water, then add a drop of liquid detergent in the centre. The soap will weaken the surface tension and pull the pepper along toward the edge of the bowl.

, a person has a thin a film of water with a mass of about half a kilogram. That's not noticeable to a human being. A wet mouse, however, has to carry its own weight plus its weight in water.

A wet fly has to carry many times its weight, so it is likely to drown. An ant or other small bug may end up trapped inside a water droplet. Surface tension makes water cling in amounts they cannot carry.

PART 3: SUMMER SIGHTS AND SOUNDS

Light and Sound

CHAPTER 19
SOUND IS VIBRATION HEARD

Tin-Cans, My True Voice, and Rocks Music

THE CHATTER of garage sales is a sound of summer. Karl and Lark, neighbours of Clay and Lacy, host the neighbourhood sale each year, combining the goods of everyone on the block into one site.

There is a lot to see. Clay and Lacy are drawn to anything retro. Last year, Clay found a lava lamp. Lacy picked up a classic wooden yo-yo. This year they found a tin can telephone, tape recorder, and ukulele.

The telephone is two soup cans at either end of a long chord. It works when the chord is pulled taut. When Lacy talks, her vocal cords vibrate to make air to vibrate. Talking into the can passes her message along the vibrating chord.

Using the old reel-to-reel tape recorder, Clay captured Lacy singing. She says the play-back doesn't sound like her, although Clay thinks it does.

It sounds better in my head.

She isn't wrong. The sound of Lacy's voice is conducted through bones in her head. Bones enhance the low frequencies and—to her—add to the fullness of her voice. The tape recording is how she sounds to everyone else.

The recording picked up shuffling and coughing—background noise we tend not to hear when listening live. The recorder's microphone does not have directionality of speech, so it picks up all ambient sounds.

In conversation, voices are directed toward one another rather than spreading out. That's a little like a megaphone. The opposite is a foghorn: sound spread all around to warn ships from any direction.

IT USED to be that every school had a set of these, Clay says, picking up the plywood ukulele and strumming across its four nylon strings. I never knew when it was noise. Like bagpipes, he adds, and they laugh.

If it has a beat, it's music.

But that could be true of banging rocks together, faster or slower, harder or softer. I could play a rock song with rocks. We will, we will rock you.

Any elastic substance (such as water, metal, crystal) can transmit sound. Rocks are elastic. They vibrate and give off a sound. Car wrenches, big and small, give off different frequencies.

CLAY

I could play a song by tossing wrenches on the garage floor if I knew the timing.

LACY

Or you could hang different size wrenches on strings and play them by tapping with a metal rod. Like musical chimes, or a kind of xylophone.

But you won't rock anyone with this, Clay says, picking up a package of plasticine. Modelling clay doesn't vibrate, so no frequency.

According to Lacy, Ajay snores in rhythm. Snore in, whistle out. One day, he might snore out the national anthem. Just as likely, says Clay, is me playing the national anthem with this. He replaces the ukulele on the table.

CHAPTER 20
SPEED AFFECTS SOUND PERCEPTION

Sonic Snap, False Falsetto, and Shower Song

ALAN ARRIVED at the garage sale with a box of items on his bicycle carrier. Look at this, said Lacy, taking a whip from the box. Totally retro.

CLAY

That the one from the carnival?

ALAN

Yes, and be careful. The tip snaps at supersonic speed, like the sonic boom of a jet.

CLAY

I can make a towel snap like that. Kids do it in gym class. Roll a damp towel along an angle, like a croissant, then snap it.

Sound is fast. It is as fast as running the full length of a soccer field, running all the way back, and running to the end again—all in one second. For the tip to snap, it has to be faster than that.

Towel snapping is one of those do-not-do-this things, like inhaling helium for a false falsetto voice. The hazard is that helium prevents oxygen from getting to the brain, so use your brain when doing it.

After inhaling helium gas, a person has a high-pitched cartoon voice since sound, such as voice, travels faster in helium. Almost three times faster. The faster the speed, the higher the pitch.

CLAY

Helium makes my voice higher; a shower makes it sound lower.

Shower tiles don't absorb sound, so Clay's voice bounces around, sounding louder and deeper. The close walls echo back his voice in sync with his singing for a rich, deep voice.

Lacy picks up an old slate chalkboard and a thick stick of chalk and she slowly slides the chalk down the board. It squeaks, like an irritating fingernail on a chalkboard.

The chalk first sticks on the board, then suddenly slips and vibrates to make the screech. As the vibrations decrease, the friction between the chalk and board increases until the chalk sticks once again. This is similar to why doors creak or the tires on a dragster squeal on a quick start.

ALAN

Chalk squeak is right up there on my irritation list, along with the buzz of mosquito wings.

CHAPTER 21

FREQUENCY VARIES WITH VOLUME

Glug Jug, Playing Pops, and Scratch Sound

As CLAY POURS soda pop out of a bottle, it makes a glug, glug, glug sound.

ALAN

Pour me some. I think the glug sound goes up as you pour out more.

Clay poured, but the pitch went lower. It isn't the pop that vibrates, but the air in the bottle. The more the air in the bottle, the lower the frequency of the glug sound. As Clay pours out soda pop, air takes its place in the bottle. More air, lower pitch.

The greater the volume of air, the lower the frequency. More air, lower glug, glug. Likewise, large organ pipes produce low notes. Some bottles start the glug, glug sound even before any liquid pours out.

Some sounds can make a person sick. That isn't likely by blowing across a pop bottle, but the drone from driving could induce car sickness. The human body contains a lot of water and sound can vibrate liquids. Frequencies below the threshold of human hearing can cause vibrations in the eyeball, lungs, and abdomen, vibrating the fluid or gas in them, causing fatigue and involuntary muscle contractions.

Lacy picks up the now-empty bottle and blows across the opening. Blowing across the edge of the bottle creates a range of frequencies. She could play a song with a row of bottles filled by varying amounts. The cavity inside enhances the resonant frequency of the bottle (determined by the size and shape of the bottle). Flutes, recorders, and organ pipes change notes by changes in cavity size and shape.

CLAY

That's like a referee's whistle.

It is. Air blown in creates a tone by vibrating an edge. The cavity shape sets the frequency. A ball bouncing inside makes the whistle warble by briefly blocking air holes.

Alan tapped the nail of an index finger against the side of the empty bottle. It made a predictable tink, tink sound. Now listen to this, he said, placing the bottle on one of the empty garage sale tables. Go to the other end of the table and put your ear to the table to listen. Clay complied. Alan again tinked the side of the glass bottle. It was much louder compared to the tink sound in open air.

In open air, sound spreads out in all directions and quickly converts to heat when absorbed. Through the wooden table, the sound does not spread out. It is confined to a smaller volume and is more directed. The sound is directed through the table like a megaphone. Or like train tracks. Sound travels faster through metal than through open air. That's why listening for a train with an ear to the rail works.

LACY

Sound travels faster through water too. Once I was in the pool, rising up out of the water as Clay lit a firecracker. I heard it underwater, then again in the air a fraction of a second later.

CHAPTER 22
FREQUENCY VARIES WITH MOTION

Radar Guns, Bat Meals, and Echograms

THE ENGINE SOUNDS LOUDER and the pitch becomes higher as Karl's motorcycle approaches. Higher because sound waves enter your ear more often. The same thing happens noticeably with a fire truck siren. As Karl zooms past, waves enter your ear less frequently and the pitch drops. Higher pitch on approach; lower pitch moving away.

Police can use this fact to measure the speed of oncoming vehicles. Police use a radar gun. Radar waves reflect off moving cars. If there is an increase in how frequently the waves that bounce back, then the vehicle is over the speed limit set by the police radar gun.

Bats also use this technique. They send out a chirp. A moving moth toward the bat, like Karl's motorcycle approaching, returns a frequency higher than the chirp. Lower frequency, if retreating.

To defend against bats, some moth species have evolved fuzzy bodies. The fuzz works like a muffler or ceiling tile; it absorbs sound—in this case, the bat's chirp—and so seems invisible to the predator bat.

If a bat and the moth were moving the same direction at the same speed, then the bat would detect no frequency change and the speedy moth would seem invisible. Same if the police are chasing a speedster.

If a large stationary ball was spinning fast, a police radar gun would detect one side moving toward it and the other half of the spinning sphere moving away. Yet the ball is stationary.

Astronomers use a technique similar to the radar gun on a spinning ball. They can tell which direction a distant star is spinning and whether the star is approaching or moving further away from us.

Lark, Karl's wife, works in the echo department of the local hospital. She uses a similar technique to harmlessly see a baby that is still in the womb using ultrasonic waves; sound higher than we can hear.

Ultrasound vibrates more than 20 000 times each second. When these waves bounce off the outside a hand or face, an image appears on a monitor. Lark can see the outline of the baby and its beating heart.

ALAN

Returning to Karl's motorcycle, the pitch changes if it is idling and I approach fast on my bicycle.

It doesn't matter who is moving closer. As long as one is approaching, the perceived pitch will be higher. That's how bats detect cave walls and stalactites.

CHAPTER 23
RADIO WAVES ARE VIBRATING ELECTRONS

Radio Tunes, Ghost Images, and Male Foghorns

ALAN HAS a transistor radio for the garage sale. Even the small ones are bulky compared to modern electronics. Sound needs vibration of matter, such as air or water. In deep space there is no matter, so no vibration. A rocket would be silent, as would a star exploding or the Big Bang itself. The Big Bang had no bang.

That said, radio waves transmit nicely in outer space. They are a different kind of wave, more like magnetism than vibration.

ALAN

Radio waves are all around us, passing right through us.

CLAY

Broadcasts from country and classical stations are passing right through me as we speak?

Yup, that and news talk, pop, and hard rock. All moving almost the speed of light.

To make the transistor radio play 101 FM, Alan turns a dial. The dial adjust the natural frequency of electronics inside the radio. When they match a broadcast signals, the station is tuned-in.

Station 101 FM sends out radio waves at a certain frequency. The electrons in the antenna of the station's transmitting tower vibrate that frequently. FM 101 means 101 million vibrations each second.

By turning the dial, Alan tunes the radio set so that the radio electronics resonate to that station. He hears music because magnets in the radio pull on paper speaker cones to match the signal.

Speakers vibrate when pulled and let go by the magnet. Vibration of the speaker cones vibrates the air and that is what we hear. Bigger and better speakers would produce better music, of course.

In an automobile equipped with AM and FM radio, the AM reception cuts out under a bridge, yet FM reception continues. AM wavelengths about the length of three soccer fields. Large objects absorb long waves.

FM waves are about the length of a ping pong table. Signals in this range are not absorbed by large objects such as buildings or bridges. Instead, they are reflected and scattered in all directions.

FM signals can become distorted. If a direct and a reflected signal from the same station are received at nearly the same moment, they cause ghost images on television and distortion or noise on FM radio.

Sound waves are absorbed as heat. Not a lot of heat, but some. Heat eventually makes its way into outer space. Higher frequencies have more energy and convert to heat faster than lower frequencies and so travel less far.

Foghorns use low frequencies, to reach further out to sea to warn ships. It is also why we tend to hear lower male voices in a crowd over the higher energy female voices at the same volume.

CHAPTER 24
COLOUR VARIES WITH LIGHT FREQUENCY

Bright White, Blue Sky, and Fast Food Red

MONA AND LARK are putting out a few paintings at the garage sale. One shows a mountain with a bluish tinge. The artist used blue to bring out the white of snow and suggest cold.

MONA

That reminds me, I want to put fabric whitener in the wash now that the kids are playing sports. A bit of bluing for their jerseys.

Blue reflects more light, making clothing appear whiter than white.

LARK

I used it on my wedding dress. My mother's dress, actually. It's an heirloom. The yellow of age disappeared. I added a little to the swimming pool and the pool has a sparkling clean gleam.

The colour seen is the hue reflected. For the dress to look yellow, the cloth absorbs all hues except yellow. The yellow is reflected back to our eyes. A black dress absorbs all hues. A white dress reflects all hues.

White is all colours combined, such as light from the sun. When white light enters a prism and exits split into all hues of the visible rainbow, red to violet. Black is the absence of colour, such as a shadow.

Fresh water and window glass are not invisible. You can see them, at least by shine and shimmer. They are colourless because they let light of all visible frequencies pass through equally.

In order to be visible, emergency vehicles are painted yellow-green. Our eyes are most sensitive to yellow-green. And the sky is blue because blue is reflected when sunlight scatters off nitrogen and oxygen in the air.

MONA

> I am not sure whether to brighten the colours of Clay's paint-ball outfit. Grunge might be an advantage. But if I don't wash them, they will have a smell-camouflage and stink.

Despite the name, there is no paint in a paintball. It contains a water-soluble dye so the colorant washes out.

A paintball is used as part of a sports arsenal, but the makeup of a paintball is more like a laxative. Most of what is inside a paintball is a chemical sometimes used to treat constipation. It causes water to be retained with the stool, but even the coating is essentially a gel capsule.

Food colours with which we are familiar are not necessarily natural. There is colouring in a hot dog, or it would be grey. Artificial colour is even added to orange juice to make it brighter orange.

Colour has associations. We associate green with the environment, but also with being raw, unripe. Green bananas, for instance. That association may be why green is seldom on milk cartons. Red and yellow makes us want to eat, so they are favourite colours in the design of fast food restaurants. On the other hand, blue and pink tend to be less appetizing. There are few blue restaurants.

CHAPTER 25

COLOUR REFLECTED IS COLOUR DETECTED

Prisms, Pigments, and Rusty Apples

UNDER THE SUMMER sun of their yard sale, Lark sets out an assortment of mismatched crystal wine glasses. Sunlight striking each goblet stem casts a red, green, blue spectrum against a box on the table. Lark turns the goblets so that red from one goblet overlaps with green from another. The result is yellow.

Lacy took crayons out of the box and drew a red patch on top a green patch. The result was brown. Adding more crayon colours eventually made black. Color from the goblets comes from the white light of the sun. Color from the crayons comes from pigments added to the wax stick of the crayon.

Comic books are printed with pigment ink in layers, building up a four colour image. Part of a scene is first printed in dots of magenta ink, then part with dots of yellow ink, then cyan ink and a bit of black.

Ink is typically used for surface application, such as printing, and contains insoluble pigments. Dye is usually soluble and used to colour materials thoroughly so that colour becomes a part of the object. For instance, a postcard is written in ink; a basket is dyed blue.

Whether ink or dye, the spectrum of light that is reflected is the apparent colour observed. If yellow is reflected, then the object appears yellow. It absorbs all hues of the visible spectrum except yellow.

Comic books, like newspapers and paperback novels, turn yellow with age, especially if left out in the sun. Comics, newspapers, and paperbacks considered disposable are printed on cheap paper.

Paper is made from cellulose, which is mostly tree pulp. Cellulose absorbs light and oxidizes (combines with oxygen). Oxidized cellulose reflects a yellow hue. The process is similar to how iron rusts.

An apple turns brown where a bite is taken. A bicycle chain rusts. Copper forms a green patina. Silver tarnishes to a dull grey. All are instances of matter mixing with oxygen and all are related to fire.

Yellowing comics, tarnishing silver, and rusting chains are slow compared to fire, but they are all related to the same process. Fire is rapid oxidation: oxygen from the air plus, say, wood at a campfire.

Rapid oxidation is why incandescent light bulbs do not contain air. Instead of air, argon gas is used inside an incandescent bulb. Oxygen in air would quickly oxidize the filament and the bulb burn out.

Nuts and seeds have a lot of oil. So do potato chips. When the oil in food oxidizes, the food has become rancid. To slow oxidation, keep food in a sealed container, stored in a cool or cold, dark place.

To keep a bag of chips from turning rancid, they are vacuum-packed in nitrogen at the factory. Air is vacuumed out and nitrogen pumped in. Nitrogen does not react; no oxidation, no rancid potato chips.

Lacy uses rancid oil, provided it does not contain salt, to lubricate her bicycle chain and keep it from rusting.

LACY

Using rancid oil to prevent rust is a slow-motion case of fighting fire with fire.

CHAPTER 26

LIGHT AND SOUND ARE FORMS OF ENERGY

Tan Ban, City Heat, and Twinkle Little Stars

ON THE ROAD A LOT, Karl drives his truck with the driver's window down, weather permitting. Arm on the window sill, he has developed a nice tan around his left elbow. The rest of Karl shows no tan.

Automobile glass, and glass in general, does not let in the kind of light that promotes tanning: ultraviolet light. Only Karl's elbow, outside the window, is exposed and so only his elbow is tanned.

Karl starts out the day with windows up, letting the morning sun warm the seats and dashboard. It is a small example of the greenhouse effect. Heat is trapped in the truck's interior.

Something similar happens in Earth's atmosphere. Water vapour, carbon dioxide, and methane gas act like the truck's compartment and trap heat. Trapped heat raises the temperature all over the globe.

The sun produces ultraviolet light, which has more energy than visible light. This extra energy can be harmful to the nucleus of a skin cell, so the body protects itself by increasing the amount of melanin in the cell.

Melanin is a dark pigment that forms a layer over the cell to filter out ultraviolet. A melanin layer is like a window shade or sunglasses. It filters light, resisting harmful rays that might damage skin cells.

Glass absorbs ultraviolet light. The thicker the glass, the more the absorption and less chance of sunburn. Clouds are transparent to ultraviolet light and so offer no protection against sunburn.

As the truck warms up, Karl tunes in his favourite radio station, FM 101. The cabin fills with light and sound. Both are forms of energy and they are absorbed by material in the truck cabin, including Karl. They are absorbed, not trapped. Trapped sound would still be in the truck's cab and could be heard if one gets close. Light and sound convert to heat when absorbed and this heat is then radiated into the atmosphere.

The radio station and ultraviolet light are on the same spectrum as wavelengths visible to Karl. Even so, Karl can see only those in a narrow band, from red to violet—the visible colour spectrum.

Most of the sunlight that reaches the Earth falls in that red to violet range. The human eye is most sensitive to light in the yellow-green range, which coincides with the peak of our sun's energy output.

The shorter the wave, the more energy it carries. Wavelengths shorter than those of visible light, such as X-rays and ultraviolet radiation, carry enough energy to destroy organic molecules in the eye.

Longer wavelengths, such as infrared radiation and microwaves, do not have enough energy to trigger changes in molecules of the retina. As a result, no visual impulses will reach the brain. No image.

Driving through the city, Karl lowers the air conditioner. City temperature is about five degrees higher than the countryside. There is less wind in a city due to taller and more complicated structures.

The city is warmer partly due to less expansive evaporation, as on a grassy field. Evaporation cools by losing latent heat going from liquid to gas. Paving and building materials also store more heat than soil does.

Karl's work often takes him to the airport. Airports are notoriously noisy. Not this one, or at least not as noisy as most. Karl's airport is on a plateau higher than the city. It is an elevated airport.

An elevated airport is quieter to the city in the valley below. Sound waves from the airport tend to go upward into the sky, rather than down into the valley. The cause is wind and temperature differences.

THE HUMAN EYE has tiny rods and cones. Rods are triggered in faint light, such as Karl driving at night. To see a faint star he looks to the side of it so the image falls on a part of his eye where rods pick it up.

We see moonlit objects mainly with rods in our eyes. Objects in moonlight appear black and white, even if they have colour by day. Light reflected by moonlight is too dim to trigger the cones, and it is cones in the eye that perceive colour.

KARL

Cruising the countryside at night, I can watch the stars twinkle.

Actually, he cannot. They don't twinkle. Uneven heating in the air makes air turbulent. The wavy air makes stars appear to twinkle.

Parcels of turbulent air bend starlight one way then another. The shimmering air can be seen above heated surfaces, such as a match or stove. Over hot pavement, it might look light water on the road.

CHAPTER 27

LIGHT AND SOUND CAN BE ABSORBED OR REFLECTED

Kitchen Prep, Knuckle Noise, and Sooty Snow

THE GARAGE SALE was a success and Karl and Lark invited a few neighbours over for a barbecue. Preparation starts in the kitchen, which, unlike the family room, is bright and noisy. The kitchen is painted white, like the exterior of the house, to keep it cooler. The kitchen is also noisy. It has marble counter tops and tile floors and sound bounces off hard surfaces such as these.

Another way to maintain house temperature is with double pane windows. Double pane has a thin layer of gas between two panes of glass. The gas is an insulator; it resists letting heat pass through.

Triple-pane windows are not much better than double since even a thin layer of gas-filled space is enough. The thickness of the gas-filled space is not significant. Triple pane is about equal to double pane.

Colour affects how much light is reflected, which affects temperature. Karl is cooler with a white shirt rather than a black shirt since white reflects better. Sunlight causes fading, so a white shirt also lasts longer.

IN THE KITCHEN, paper towel is used to absorb spilled milk. Paper fibres have small spaces to take in the liquid. Absorption of sound also needs small pockets. Hard surfaces will reflect sound back in a room.

Aluminum foil is smooth, at least on one side. It is dull on the other side. That is a result of how aluminum is made. Both sides are the same metal and so, for baking, it does not matter which side is in or out.

The family room has carpet, drapes, a fabric couch and acoustic ceiling tiles, all of which trap light and sound. Acoustic tiles act like an automobile muffler: sound fills small chambers, called baffles.

Spilled milk can be absorbed with paper towel. Gases can also be absorbed, sometimes even into liquids such as the fizz in a can of soda pop. Carbon dioxide absorbed in pop gives it fizz.

Knuckle cracking is like pop can fizz. There is fluid between finger joints, synovial fluid, that absorbs carbon dioxide. Pulling or pressing makes knuckles "crack" by separating bones, lowering pressure.

Lowering pressure releases the absorbed carbon dioxide as bubbles. As bubbles collapse, they produce a pop sound. After a while, the gas reabsorbs into synovial fluid and knuckles are ready to crack again.

Light rays can also be absorbed. Garden dirt absorbs most of the light that hits it, making it a poor mirror. Albedo is the percentage of energy reflected from an object. The albedo garden soil is about 10%.

To say the albedo of soil is about 10% means dirt reflects 10% and takes in 90%. Higher albedo means higher reflection. Fresh-fallen snow is about the opposite of soil. It reflects about 90%.

To prevent floods, airplanes drop soot on snowy mountain sides before spring. Soot has a low albedo; it absorbs energy from the sun and so helps the snow melt slowly to prevent rapid-melt run-offs.

Light on an object will be absorbed (taken in), transmitted (passed through), or reflected (bounced back). The charcoal briquettes in the BBQ absorb light. They are opaque. You can't see through them.

A glass pitcher of water is transparent: you can see right through it. Waxed paper, under the burger patties, is translucent. It lets diffuse or scattered light pass through, like a frosted-glass shower door.

PART 4: ECLECTIC AUTUMN

Electricity and Entropy

CHAPTER 28
STATIC ELECTRICITY IS BUILT-UP CHARGE

Fall Fashions, Static Spark, and Dust Magnets

MAPLE LANE, where neighbours hold their community garage sale, is so named for the many sugar maples and other colourful deciduous trees. This autumn, with strong sunlight and cool dry nights, the leaves are brilliant red.

After summer solstice, with less light each day, the maples pull nutrients back into their trunk and roots. The leaf stops producing the green pigment chlorophyll, revealing splotches of red and yellow.

The cool, dry weather is also ideal for static electricity, both big as lightning and small as sparks. Friction between shoes and a nylon carpet can cause static electricity to build up on a person, although static spark is less likely on humid days. Moisture in the air neutralizes the charge.

Static electricity makes clothes cling and produces in a small spark when metal is touched. This spark or discharge is, in fact, miniature lightning. That is, finger sparks and lightning are static electricity.

Clouds become charged as ice crystals inside them rub against each other. The build-up eventually cannot be held in the cloud and a flow along the path of least resistance is found as a bolt of lightning.

Static sparks can occur between clothes coming out of the dryer, or by taking off a sweater on a dry day. Static cling can also hold a balloon to a wall after the balloon is rubbed on a person's hair.

Static electricity can be harmful to sensitive electronic equipment, such as repairing a computer circuit. It can be dangerous during surgery if a heart or other sensitive organ receives the spark. In fact, trucks carrying flammable cargo used to drag chains to discharge static electricity built up by friction with the air. Now, the tires for trucks carrying flammables are made with metal strips built-in.

Lightening that hits the ground can travel up one leg and down the other, leaving a person temporarily stunned. The further apart the legs, such as with a cow, the greater the amount of current across the body. Lightning that strikes near a house can even enter through the plumbing system, so it is best to wait on a bath until the storm passes.

During a lightning storm, don't stand under a tree or touch an automobile. However, it is safe to be inside an automobile. The metal body conducts electricity, so the current stays on the outside. A metal airplane likewise protects passengers once aboard the plane.

Static electricity is useful in spray-painting metal objects, such as auto parts or patio furniture. The positive paint, attracted to the negative metal like a magnet, covers in one coat with no dripping.

A photocopier is similar. Parts of a rotating drum are charged to attract the powdered ink toner. Paper passes through, picks up the toner by static cling, then heat melts the toner into the paper.

An electrostatic air purifier is a magnet for dust in the air. Particles flowing through the filter are given an electric charge. They are then attracted to the magnetic plates in the purifier.

CHAPTER 29

LIKES POLES REPEL; OPPOSITES ATTRACT

Pole to Pole, Iron Fortified, and Cooking by Friction

IRON IS in bridges and buildings and blood. Iron helps carry oxygen in red blood cells, for instance. Some breakfast cereals are fortified with small bits of iron. Breakfast for Clay is often a bowl of cereal, packed with vitamins and fortified with iron.

CLAY

A magnet will attract iron. If iron is in my cereal, my cereal must be magnetic.

The iron in dry breakfast cereal is the same mineral as in a cast iron skillet. In the digestive tract, iron changes into a form the body can absorb and use. A strong magnet can pull the iron out of cereal fortified with iron.

Magnets have a north pole and a south pole. Break a magnet in half and each half becomes a magnet. The Earth is a magnet, with north and south poles. An atom is a magnet: negative electron, positive nucleus.

Opposites attract. The north pole of one magnet will attract the south pole of another magnet. But a magnet will also attract metal that is not a magnet. Iron, for instance. Both poles will attract the iron.

A magnet sticks to iron by making it a magnet. The north pole makes an iron bar temporarily into a south pole and attracts it. The south pole makes it temporarily into a north pole and attracts it.

Some animals use magnetism to navigate. Pigeons and Monarch butterflies, for instance, produce grains of iron magnetite. They use magnetite with the Earth's magnetic field to navigate, like a tiny GPS.

Opposites do attract, yet north on a compass points north. That's because north on a compass is really the south pole part of a compass needle, painted to look like it is showing Earth's north pole.

WATER IS HYDROGEN AND OXYGEN. H_2O. A water molecule is a small bar magnet: negative on the oxygen side; positive on the hydrogen side. A microwave uses water like bar magnets to warm food by friction.

A microwave oven sends a magnetic wave into the box where food is placed. The water molecules in the food flip to line up with this magnetic field. The wave, however, keeps flipping back and forth.

As the magnetic field flips, water molecules in the food flip to keep up. Two and a half billion times every second. Food heats up by friction between flipping water molecules and the rest of the food.

Paper and glass have little water, so they don't heat up in a microwave oven. Popcorn has moisture in the kernel that heats up and busts open, sending a sound wave with a characteristic pop, pop sound.

An ice cube has a lot of water, but frozen—fixed in place. As a result, a microwave oven does a poor job heating up ice. Metal reflects microwaves, which could cause damage or even a fire in the oven.

CHAPTER 30

MAGNETIC ATTRACTION HAS PRACTICAL USES

Magnetic Images, Household Uses, and Finding Poles

LARK WORKS in the imaging department of the local hospital. One device in her department is the MRI or magnetic resonance imaging machine. A patient lies in a tube and surrounding magnets align water molecules in the patient's body. Radio waves then pick up signals and create detailed images in slices, like slices of bread.

At home, Lark finds magnets all around. In cabinet door latches and audio speakers. On credit card stripes and purse snap closures. In motorized devices from electric toothbrush, to cordless drill, to garage door opener.

Karl has a variety of uses for magnets in the garage, especially the strong neodymium magnets. Using a hot glue gun, Karl stuck disc magnets on the wall. Now they hold up pliers, hammers, clippers, or any metal tool. He puts one neodymium magnet on a screwdriver so it will pick up and hold screws in place. It works. With another magnet is his shirt pocket, screws and nails stick to the outside of his shirt for easy access.

When Karl spilled a box of nails, he held a neodymium magnet on the bottom of a cardboard box and turned the box upside-down over the spilled nails. The nails jumped up, back into the box.

To clean the inside of a fish tank, Lark borrowed an idea from the surgery team. Surgeons use powerful magnets outside a patient's body to guide instruments and perform procedures with fewer incisions.

Lark put a neodymium magnet in a sponge and put the sponge into the fish tank. Using another magnet on the outside, she moved the sponge around, scrubbing the tank without disturbing the fish.

Doing maintenance work, Karl developed an appreciation for the properties of shapes. Maintenance hole (manhole) covers are round, for instance. That is so they can't be accidentally dropped into the hole.

A torus (donut shape) could be turned into a single-handled coffee cup without tearing or punching. They both already have a hole. A cube could be turned into a cylinder. But a cube can not be turned into a torus without tearing or punching.

So a horseshoe magnet is equivalent to a bar magnet by shape and by having north and south poles. Both are equivalent to a spherical magnet. Lark picks up an orange and tries to imagine it as a magnet.

Think of it this way: when the wind is blowing, there must be another place on the globe where the air is still.

Magnetic flow around a sphere also has an origin and destination: the poles. Every magnet has two poles. Even a sphere like the Earth has two magnetic poles. Put a couple magnetic spheres together and they will align to attract north to south poles. Sphere magnets stack easily.

CHAPTER 31
ELECTRICITY FLOWS THROUGH A CONDUCTOR

Electro-Quiz, Brain Sparks, and Ohm Lying

Home-made gifts are a tradition in Mona and Noam's family. With the holiday season coming up, Lacy made a family-history quiz for her grandmother. First, she wrote questions on the front of a manila folder.

Lacy taped aluminum foil strips inside, connecting choices with answers, punching holes beside each. She put paper clips at both ends of a length of bell running along a small bulb and battery.

Touching one paperclip to a question and the other paperclip to the correct answer completes a circuit and the bulb lights up. Of course the manila folder is taped closed, so grandma can't see inside.

With a correct question to answer connection, the bulb lights up. It lights up right away, although electrons move through the copper wire about as fast as an ant walks, one centimetre each second.

Electrons move at the speed of an ant, but the pulse of an electrical signal moves along the wire at nearly the speed of light. In one second, the pulse could go around the Earth's equator seven and a half times.

Paper is an insulator. Lacy put paper tape between the foil strips to prevent short circuits. In a short circuit, the electrical pulse would take the shortest route rather than the intended route.

The battery is rechargeable, but it seems to take a little longer each time to recharge. It's that way with the battery in her smartphone and electronic tablet. Fast to partially charge; slow to fully recharge.

Rechargeable batteries charge quickly at first, but take much longer to finish fully charge. They are like suitcases in that sense. The fuller it becomes, the harder it is to add more and the longer it takes.

As Grandma tries out a lot of different choices, the wire becomes warm. Most of the electrical resistance is in the wire. Power lost due to resistance is converted into heat, such as the coils in a toaster.

The longer the wire and thinner it is, the more the resistance. It is like watering the garden through a long straw. The water has a harder time passing through. Shorter and wider, less resistance, like a fire hose.

Lacy's brain is a bit like this. Impulses travel along nerves as an electrical wave, although they ferry across from one nerve to another by means of chemicals. Her brain produces 10 to 15 watts of power. That's about as much as a single LED light bulb uses.

∼

A LIE DETECTOR is similar to the manila folder quiz, but with sweat for aluminum strips. When a small voltage is applied, a sensitive meter shows whether current flows across the skin when a person sweats.

Lying promotes perspiration. Perspiration consists of a salty water solution, which is a good conductor. Better conduction (lower skin resistance) results in a higher meter reading.

Lie detectors aren't very reliable, however. Some liars don't sweat under pressure and other people telling the truth do. Even nervousness of taking a lie-detector test can give a false reading.

A lie detector used by some poker players is dilation of the pupil. Look into the eyes of an opponent to detect a bluff. Contracting pupils suggest that a person is displeased with what he or she senses.

CHAPTER 32
AS ENERGY CHANGES, IT BECOMES LESS USABLE

Magnet Train, Heat Change, and Things Deteriorate

As a present for Clay, Lacy made a train out of neodymium magnets, a AA battery, and bare cooper wire. She tightly wound a length of the wire around a wooden dowel to make a long coil a little larger in diameter than the diameter of the battery.

The magnets are discs, the same diameter as the battery. Lacy put one magnet on each end of the battery, north poles touching the battery. The train is made of disc+BATTERY+disc.

Put the train inside the coil and it immediately begins to moves along inside the coil from. It moves from one end to the other, or will go around and around without stopping if the coil is bent into a circle.

The magnets touch the coil and create a circuit from one end of the battery to the other. This creates a magnetic field around the wire that repels a magnet on one end and attracts a magnet on the other.

Electricity flows in the circuit, which means the train will keep moving until the battery runs out. The AA battery runs out when the chemicals inside break down, unless it is a rechargeable battery.

∼

A BATTERY RECHARGER reverses the flow of energy using a power source, such as solar panels. Energy cannot be created or destroyed, only change form. A solar panel changes light energy into electrical energy.

Household electricity comes from the flow of water through a dam. Water turns a turbine, which generates electricity. Coal, wind, and steam can also be used to turn the turbines that generate electricity.

Things deteriorate, becoming less usable, less organized. Batteries run out. Lacy's room goes from clean and tidy to a bit messier each unless put back in order. A car runs out of gas. A campfire dies out.

To make it usable again or restore order, energy has to be put into a system. The battery has to be recharged; Lacy has to clean up her room; the car has to be filled up; the camp-fire needs more wood.

Heat flows from warmer to colder unless work is done to reverse it. For instance, an air conditioner to offset summer heat. A furnace in winter to replace heat lost through windows and doors.

The universe as a whole is like Lacy's room. Energy can be converted, but some is lost as waste heat. The amount of disorder or waste heat builds up. It is not possible to get more energy out of a system than is put in.

As Ajay knows from his cabin, you can't cool off a room by leaving the 'fridge door open. More heat is given off by the motor than is absorbed by the released cool air; the room becomes even warmer.

CHAPTER 33
AN ELECTRIC MOTOR USES MAGNETISM

Mini Mags, Rowing the Boat, and Making a Motor

LARK BOUGHT a strip of flexible magnetic tape for crafts and fridge magnets. She cut it in half. Both halves have north and south poles and are magnets. Cutting them in half, she has four magnets; cutting again, she has eight.

Using a magnifying lens, she kept cutting until the bits became invisible to the eye. Invisible, but still magnets. She could keep cutting until there are only atoms or only bare electrons. Still magnets.

Electrons are the tiniest magnets. If a lot of them spin the same way, the material is magnetic. Iron isn't magnetic and that's because some electrons spin this way, some spin that way, and so they cancel out.

Electrons canceling is like one person rowing forward while another rows backward. The boat stays put. If only one person rows or if two row one way and a third rows the other, then the boat moves.

The boat moves as long as there is an unpaired rower. That is, someone rowing one way not paired with someone rowing the other way. Electrons are like that. All it takes is one extra going one way.

A material that has unpaired electrons is magnetic; it can become a magnet. Iron, for instance. A material that has paired electrons is non-magnetic. Copper is an example.

A steel bolt isn't magnetic. Steel is mostly iron, as much as 99% iron. But put the steel bolt near a magnet and patches of electrons called domains align in the same direction and the bolt becomes magnetic.

A magnet will stick to another magnet because opposites attract. A magnet will stick to a steel bolt because it magnetizes the bolt. A magnet won't stick to a stick of wood because wood has no domains.

Another way to make a magnet is to run electric current through wire. The magnetic field is invisible, but a compass near a wire that is part of a circuit will be affected by the magnetic field through the wire.

An ELECTROMAGNET IS the basis for an electric motor. Lay a battery on its side and balance a disc magnet on the battery. Tape a paperclip to one end of the battery. Another to the other end of the battery.

Loops uncoated copper wire around and around into a coil of many loops, but leave the ends sticking out from the coil like –**O**– and insert the ends like arms into the paperclips. The coil will start spinning.

The sticking-out ends are a miniature drive shaft. This is the basic motor for everything from a toothbrush to a garage door opener. The battery creates electrical current and magnetic field in the wire.

The loop becomes an electromagnet. It is first attracted to and then repelled by the disc magnet on the battery. Attracted, repelled, Attracted, repelled. Around and around it spins, turning the drive shaft.

CHAPTER 34
THINGS WEAR DOWN UNLESS KEPT UP

Messy Party, Magnetic Slime, and Crumbling Cookies

WHILE MAKING HOME-MADE GIFTS, Lacy discovered that things deteriorate. Batteries run down, unless recharged. Her room gets messier, unless cleaned up. That's entropy for you: the tendency of things to become more disorganized, less useful.

LACY

Perhaps entropy is why the classroom becomes chaotic when the teacher leaves the room.

Entropy is at work at a party. At first, the chairs are all lined up. Gifts, neatly wrapped. Clothing pressed and food ready to serve. As the party wears on, there is more to clean-up, tidy-up tomorrow.

Lacy wants a gift for Alan, ideally one that does not become more disorganized. Perhaps a jig-saw puzzle. Out of the box, the pieces are jumbled up. Putting the picture together, there is less and less chaos.

That is, unless the bigger picture is taken into consideration. The food consumed by those putting it together. The gasoline used to shop for food. Electricity for lighting and energy to make the cardboard.

Life itself seems to wear down considering the resources needed going from a baby to senior citizen. So when thinking of entropy, Lacy needs to consider all that goes into making and playing the jig-saw puzzle.

Lacy decided to make magnetic slime. She mixed white glue (the kind used in school), water, green food colouring, baking soda, and iron filings. It mixes into a putty that does not appear organized or useful.

Magnetic slime sits there like a snotty blob—until near a magnet. The blob will ooze along to follow the magnet and, if close enough, snap out to catch the magnet like a frog catches a fly. Drop on a refrigerator magnet and the blog wraps around, swallowing up the magnet like a sci-fi creature. In the presence of the magnet, the blob seems to become more organized and capable of movement.

Cleaning up, it occurs to Lacy that the blob is like the jig-saw puzzle, given the effort and resources that go into it. That's when she bumped her mug of cocoa and the mug shattered on the kitchen's tile floor.

LACY TRIED PUTTING the pieces back together, like Humpty Dumpty. Forces between molecules hold a cup together, so putting pieces together carefully they should stick by attraction between molecules.

It doesn't work. To adhere, surfaces must be unimaginably close, about the width of an atom. Solid surfaces are too rough to allow more than a tiny amount of their surface area to be this close.

The liquid glue used to make the magnetic blob worked better. Most adhesives are liquid, at least initially, because of the need for close contact between the glue and the materials being stuck together.

A liquid glue can flow into surface irregularities to provide the needed close contact. Close contact is also prevented by settling dust particles and distortion of the cup during the break.

Unfortunately, white glue does not form strong bonds and Lacy relegated the mug to the trash. She poured a glass of milk and took out a cookie. No hot cocoa, so cookie dunked in milk would do.

Dunking draws the liquid into pores in the cookie and dissolves the sugar structure of the cookie. The cookie falls apart. The longer the cookie is submerged, the weaker it becomes. Structures deteriorate.

CHAPTER 35

RESISTANCE ACTS TO OPPOSE FLOW

Fresh Toast, a Babbling Brook, and Perpetual Motion

THE FAMILY-HISTORY QUIZ Lacy made for her grandmother uses a simple circuit. Connecting a question with the correct answer, completes the circuit and a small bulb along the circuit lights up.

To light up, the bulb is incandescent. Inside the bulb is a thin wire coiled up to create resistance. If there is enough resistance, the wire will heat up and produce light. A little light; a lot of heat.

Water flows through a hose. If the hose is long and thin, that makes the flow more difficult. Electric current passes along a conducting wire. If the wire is long and thin, it too resists flow.

Incandescent bulbs are usually made with a long, thin coil of tungsten in a bulb of argon or nitrogen gas. Tungsten resists corrosion. Inert gas prevents the filament from burning out as it would in oxygen.

Incandescent lightbulbs give light. Mostly they give heat, about 95%. Heat from a light bulb is wasted and again shows that things deteriorate, becoming less useful.

SOME APPLIANCES MAKE use of resistance through a wire when heat is wanted more than light. A water hater, for instance, or hair dryer or a toaster. In such cases, it is not light but heat that is useful.

The wire in a toaster is made of nickel and chromium. It can become hot, but does not rust or melt. When hot, it gives off visible light and appears red, but toasts the bread by infra-red heat.

Toasting is not the same as burning bread, turning it to carbon like the ash from a fireplace. Toasting causes sugar and amino acids (which make protein) in the bread to change the texture and flavour.

The same process occurs when barbecuing, roasting, or baking. Going from raw meat to steak well-done. From uncooked potato to crisply roasted. From pie dough to a golden crust.

RESISTANCE OCCURS in water circuits as well, similar to an electrical circuit. For water, the source is a pump. For electricity, a battery. They are connected by a pipe or wire; controlled by a valve or switch.

Lacy made her mom a miniature babbling brook for relaxing white noise at the end of a busy day. The waterfall sound is similar to the toaster's heat in that both are the result of resistance along a circuit.

She bought a fish tank at the garage sale. A pump at the bottom of the half-filled tank draws water in from one side and pushes it up a tube, along a small diving board, and onto rocks and pool below.

The water flow encounters some resistance through the tube, but in the rocks as well. Were there no resistance and a perfectly smooth circuit of water, there would no babble to the brook.

The water falling into the tank turns a small fan. As the blades turn, they rotate a coil between magnets and produce an electrical current—basically, the same way electricity is generated at a hydroelectric dam.

Falling water or steam or even pedal power can be used to turn a coil between magnets and generate electricity. As Lacy pedals her bicycle, the tire turns a coil in a generator that powers the bicycle headlight.

The water pump generator helps power the pump, but it alone is not enough to power the pump since there is resistance and lost heat. If it could, it would be a perpetual motional machine and defy entropy.

CHAPTER 36
MEASUREMENT AFFECTS WHAT IS MEASURED

Pumped Tires, Active Fans, and Uncertainty in Photography

Lacy's bicycle, like most bikes, has pneumatic tires. They have to be pumped up. There is a recommended pressure imprinted on the side of the tire and she pumps the tire to the maximum and a little more. Over-filling makes the tire more rounded for less contact with the road. Less contact means less resistance due to friction, and her peddling becomes more efficient.

Although Lacy pumps each tire to the max, when she removes the pump head, air escapes. Sometimes a little and, when she has more difficulty disengaging the pump head, sometimes a lot.

As a result of escaping air, Lacy never knows exactly what pressure is in the tires. Even over pumping results in a best estimate, not accurate measurement. There is a certain amount of uncertainty.

Lacy is a cub reporter for the town newspaper and covers baseball. She conducts interviews and moves about the ballpark taking pictures, often catching the eye of local or visiting teams. Even a slight distraction can have an impact, such as a player distracted by the flash of a camera. At times Lacy affects the outcome in the process of reporting: news reporter turned newsmaker.

It is not possible to accurately observe or measure that which is modified by the process of observing or measuring. Reporting that changes what is observed is making news rather than reporting news.

The crowd can get into the game with enthusiastic cheering and chanting, inspiring the players. Even those watching the game televised feel their cheering can have an impact on the outcome. Except there is no feedback loop from those watching on television to the players. However enthusiastic the cheering at a local pub, viewers have no impact. They are passive.

It might make a difference if the players knew how many fans were watching the game televised. More viewers, more enthusiasm to perform which, in turn, might impact the outcome of the game.

Clay pitched a fast ball and Alan hit it out of the park. Lacy's photo, with a slow shutter speed, showed a blurry image of the ball. It gave a sense of speed and direction, but not the ball's exact location.

Ajay's camera, with a fast shutter, shows the baseball in focus. Its exact location can be determined from the background, but the ball appears suspended, with no indication of speed or direction it was headed.

EPILOGUE: BEYOND THE YEAR

Ideas on the Whole

CHAPTER 37
NATURE IS NEVER ENDING

Space Pulse, Many Worlds, and Billiard Balls

LET's start at the beginning. The beginning of time. Start with the Big Bang that wasn't big and wasn't a bang. The start of it all was compacted so small as to be indescribable. For reason's unknown at this moment in the story, it blew up. Rather, blew out.

It blew outward and continues to expand (except it doesn't and more on that later as well). Since there was no air or other medium for sound, there was no bang. In space, nobody can hear you explode.

The smallest of bits blew outward, but the centre point remains a mystery. Maybe one day we will find the point from which everything is "out." One day everything will blow back in. More on that later too.

I ASSERT that the tiniest of bits have a property as basic as being. To be is to have this property. For the sake of simplicity, and this is a story of simplicity, that property is attraction. To be is to attract. Even as all the smallest of particles were flying apart, they were drawn together.

Attraction isn't quite gravity. Gravity is a result of the basic attraction of matter to other matter. Attraction isn't a property of matter—and this is important. It is another way of looking at matter. Not the flip side of the coin, but one and the same as the coin. To be is to attract. To attract is to have being.

Particles combined. In those two words are found the origin of time and the universe. Let's start with the universe. Time doesn't actually exist. At least, not on its own.

THE COMBINATION of particles settled in levels. The atomic, molecular, familiar, and cosmic. At the atomic and subatomic level are atoms and their parts, such as electrons and protons. But also bosons and quarks. Subatomic could be put in its own level, but they are all so horribly small that the distinction bears no difference to us. These are the billiard balls of the universe.

How much bigger is the molecular world? Measured size isn't as important as awareness. That is going to take a bit to understand, but as said: it is important. Big ideas sometimes need big words. **Epi-** (which means above or beyond) **phenomenon** (which means occurrence, something observed). We will use this big word in a moment.

Water is a molecule. So is table salt. Propane for the BBQ is a molecule, so is chlorine in the swimming pool. Molecules arise out of atoms, but occur above or beyond atoms. Take the simplest: water. Water is two hydrogen atoms and one oxygen atom. H_2O.

In the world of water (which is the molecular level), there is no observable hydrogen (which is in the atomic level). There are no electrons or quarks in the observable world of water. If you could ride a tiny ship from the world of water down to the world of atoms, water would disappear.

In other words, water is an epiphenomenon of hydrogen and oxygen. Hydrogen consists of a single proton and single electron. If we look around the atomic world, we see protons and electrons, but no water. No salt, no propane, no chlorine.

So we can go up observable levels, from atomic to molecular or down them, molecular to atomic. And up from molecular to cellular. Skin cells, tree cells. They arise from molecules, but under the microscope, no molecules appear. They are just a nucleus and cytoplasm and other bits.

We are made of cells, but no matter how closely I look, I see no cells in you. That's because people (familiar level) and cells are on different levels.

Cells can be their own level, if you want, somewhere between molecular and familiar. The edges become blurry. For instance, the largest cell is observable with the naked eye. Just barely. This is the oocyte, or human ovum. The oocyte is colossal compared to most cells.

Pictures in newspapers are made of tiny dots of ink. If you put nose to newsprint, you see only a lot of dots, no picture. Moving away from the newspaper, an image appears. Abraham Lincoln, for instance. Up close, there is no observable image. Further away, there are no observable dots. The image and ink dots occur in different levels of observation. The image of Abraham Lincoln is an epiphenomenon of ink dots.

Our familiar level consists of trees and rocks and animals and oceans. Oceans are on the edge. Nobody can behold an ocean as a whole. Certainly not the world as a whole and much less a solar system or galaxy. Let's put these in the cosmic level.

If we became as big as a galaxy, we would see no people. If we became as small as a molecule, we would see no people. Galaxies, people, and molecules occur in different levels of observation. So however you stratify or divide up the layers (atomic, molecular, cellular, and so on), what matters is whether what is observable as a whole in one level is observable in another. And, for our purposes, the atomic level is the foundation upon which all other levels rest. That is important because atoms are not unlike billiard balls.

First a caveat: atoms do not look or act like solar systems. An electron does not orbit a nucleus in the way a planet orbits a star. An electron has a negative charge that creates a force field. Force fields can be very strong. Most of matter is empty, almost entirely empty, but you can't walk through a wall by jiggling your atoms. You can't pass your feet through the floor. The wall and floor are mostly empty space, but and this is a big BUT, electro-magnetic force fields make matter seem sold.

Now lets take back that caveat and simplify matters. Atoms are pretty much like billiard balls. Imagine a set of them racked up on a pool table and break them sharply with a cue stick. When the cue ball strikes, the other balls scatter. Action, reaction. The solid and stripe balls, have no choice. They don't think about it, but obey. At the atomic level, there is no free will. Yes, atoms are not billiard balls. This is a simplification. But still, they have no choice. At the atomic level, there is no free will. How the balls move is determined by how they were struck. Billiard balls and atoms are determined.

The BIG question is whether determinism scales. That is, because atoms have no choice, do molecules have a choice, do people have a choice, do galaxies have a choice? It seems to us that we do, but seems does not entail is. Physical change is determined, such as the cause and effect of cue ball and billiard balls. Mathematical change is determined, such as adding two plus two. Always four.

I argue that there is no reason to believe the familiar world has free will when all other worlds are determined. It seems to us that we have freedom from cause and effect. The key word being: seems. But if mind states are an epiphenomenon of brain states, then your thoughts this moment are determined by thoughts before and any sensory input in the moment. Brains are made of cells are made of molecules are made of atoms. And atoms act and react like billiard balls.

And if mind states are not an epiphenomenon of brain states, then there is an unexplainable ghost in the machine. To say minds were put there by an outside force is to slide down the slippery slope of who put a mind in the outside force and who put a mind in the mind of that force. On and on, unless one short-circuits the process and declares: just because. Let's hope for more of an explanation than: just because.

Earlier I said: in the words "particles combined" are found the origin of time and the universe. It turns out there is more than one universe of awareness. There are stacked universes: atomic, molecular, familiar, cosmic. We will return to these after we dispose of time.

We left the familiar universe expanding rapidly after the now unfortunately-named big bang. It was, simply said, stuff in motion. Stuff that became bosons and photons, neutrons and protons, eventually crayons and drive-ins. But at first it was just stuff. Still is stuff, but the kind that is familiar to us. The stuff of our familiar level of epiphenomena.

Like objects moving relative to one another. Earth rotating about its axis. The flight of Wilbur Wright. The fourth and final flight of the Wright brothers lasted as long as it takes the Earth to rotate on its axis, or at least 1/1440 rotation.

There are objects in motion: Wilbur flying; Earth rotating. Motion can be compared, one against the other: the final flight of Wilbur right takes 1/1440 rotation of the Earth on it's axis.

We can call this a minute, except time does not exist apart from relative motion. There are objects, there is motion, there is comparison. But there is no time. Time is a short-hand that becomes inconvenient when a minute or second or eon is thought of as existing in its own right. There is no time, only relative motion.

Time does not exist apart from matter. No matter, no motion. No motion, no comparison. No apparent time. In an absolute vacuum, there is no time.

Before the big bang, there was no time. Time (relative motion) began only when it all blew outward. And by it, we mean matter of whatever form, including the photon, which moves as a wave and strikes as a particle. Here we return to the interpretation of matter as attraction. It all blew apart and the parts have been (figuratively speaking) calling out to one another ever since.

One theory is they will smear out into a heat death and the universe stay like that, an unmoving soup where time (relative motion) slides to a halt when there is no further motion.

However thin the cosmic soup, to be is to attract—and attract the bits of the universe will. Collecting, aggregating, pulling in on itself, tighter and smaller until a single point occurs. A singularity which blows apart and the universe pulsates out, then in, then out.

But not randomly. There is no random; only a hard determinism at the atomic level. What was, is now, and will be again. Over and over, world without end.

You were here before. Not the here of this instance of the universe, but here in the instance before, the big bang blowout before the universe crunched together and blew out again into this world. You will be here again. And because the universe plays billiards, you will do as you did.

It isn't like you have to wait long. Imagine a dreamless sleep. You fall asleep one moment and wake the next. This great cycle between one instance of you in this universe and you in the next could be written as a number with a large exponent, but what does that matter. You blink out, blink back in.

Not the same you; not a reincarnation. It is a you that comes about because the stuff of the universe reaches the point where the right combination of matter creates just that brain state that gives rise to the mind state that you know to be you. You are inevitable. You are eternal. You just don't know it. Now you do.

You can't know anything about the previous instantiations. They are discrete because all matter smoothed out into a cosmic soup and crunched in on itself and blew out again until there you are, thanks to hard determinism.

So the universe pulsates: out, in, out, in. And each time, eventually, stuff comes together in a composition that is uniquely you and there you are. Out, in. Out, in.

How many times have you been before? The question is simple, but it makes no sense. There was no Universe-Zero. That's true in part because there is no nothingness.

We talk as if nothing exists. That is, as if there is a nothingness. There isn't. Energy cannot be created or destroyed, only converted from one form to another. Something cannot come out of nothing, because nothingness does not exist.

We can think of a vacuum, of a patch of void in deep space where there are no atoms. That is emptiness, not nothingness. We can think of contrast, such as a shadow, where there is an absence light. Or hole where there is no donut. Or cold, which is the absence of heat. But these are apparent by contrast. A shadow, or donut hole or cold are a description of what is out there, but not here at a location. There is light, but it is dark here. There is heat, but it is cold here.

That there is no nothingness entails that what is always was. Nothing can come from nothingness, so what is always was. In short, no creation—only outward expansion from the singularity that started with a big bang.

The stuff of this instance of the universe is the same stuff of every instance of the universe. The stuff in you is the same as it has always been and always will be. You were merely waiting for the billiard balls to reach the configuration where they came together as your brain and, with it, your mind, your conscious sense of self.

You have been instantiated not a number of times—but always. That is because the stuff of the universe pulsates out and in, out and in, never creating new, never losing as much as an atom to nothingness. And you are a small part of that uncreated, never-ending stuff across all instances of the universe. You always are, were, will be. Everything is fixed and you can't change it.

CHAPTER 38

NATURE IS EVER EMERGING

Beasts, Herds, and Self-Awareness

The ground beneath you and the sky above,
beasts of the forest and the chillin' thereof.
Atoms and planets and the stars at night. If it
walks or swims or even takes flight.

It appears as whole, it seems to be one. More
than its parts, yet greater than none.
Similar bits are what nature favours. They
interact, at least with their neighbours.

Geese will gather and fly in formation. Planets
form by self-organization.
Towns arise from economic exchange.
Diamonds are crystals so neatly arranged.

Internal harmonic brings parts into sync, like
fireflies when the whole field blinks.

Or human cycles in tune with each other by
subtle signals we've yet to discover.

Hives, herds, swarms, flocks, schools of fish.
Emerging in turn, to flounder or flourish.
Emerging to represent internally, that which we
all experience externally.

Patterns inside that roughly correspond to their
world within and the world beyond.
Experiencing experience, itself self-aware,
conscious of being and being somewhere.

Greater complexity, finer in clarity. Deeper in
awareness, higher in verity.
Able to label and categorize. Construe it, intuit,
and rationalize.

The mind of an ant at the hiker's boot is how the
hiker stands in Gaia's pursuit.
From Gaia to galaxy or structures we can't see,
conscious beyond any descriptive degree.

To the universe itself and invention of time.
Relative motion, yet each movement's
prime.
What is always was, there never was nothing. I
think, I am; always was, yet becoming.

ABOUT THE AUTHOR

Previously published **Pale Blue Lives**, a collection of fables, parables, tales, and verse. When not writing, I'm riding—eBike, motor bike, and a mow cart that catches air down the hills. I have a companion site on cookiejaropen.com and sister site on zipwits.com. One day I'll have Goldies again.

WONDERMENT

Science in Stories

Deceptively simple, Wonderment makes postcards of big ideas. It visualizes through humour and examples, rather than burying big ideas under maths and charts. This might be how Hemingway would have taught science, in stories. Say as much as needed and leave the reader to fill in from experience.

In all, Wonderment has a way of making connections that are only obvious once you know them. And it progresses this way from how the parts work to a speculative essay on the whole, from physics to a metaphysics, without losing its simple wonder of the way the world works.

www.ingramcontent.com/pod-product-compliance
Lightning Source LLC
Chambersburg PA
CBHW051422130726
47989CB00014B/634